ORIGO

# STEPPING STONES

CORE MATHEMATICS

**SENIOR AUTHORS**

James Burnett

Calvin Irons

**PROGRAM CONSULTANTS**

Diana Lambdin

Frank Lester, Jr.

Kit Norris

**CONTRIBUTING AUTHORS**

Debi DePaul

Peter Stowasser

Allan Turton

**PROGRAM EDITORS**

James Burnett

Beth Lewis

Donna Richards


ORIGO EDUCATION

**PRACTICE BOOK**

# INTRODUCTION

## ORIGO STEPPING STONES

The *ORIGO Stepping Stones* program has been created to provide a smarter way to teach and learn mathematics. It has been developed by a team of experts to provide a world-class math program.

### PRACTICE BOOK

Regular and meaningful practice is a hallmark of *ORIGO Stepping Stones*. Each module in this book has pages that practice content previously learned to maintain concepts and skills, and pages that practice computation to promote fluency.

**PERFORATED PAGES**
The pages of this book have been perforated for your convenience.

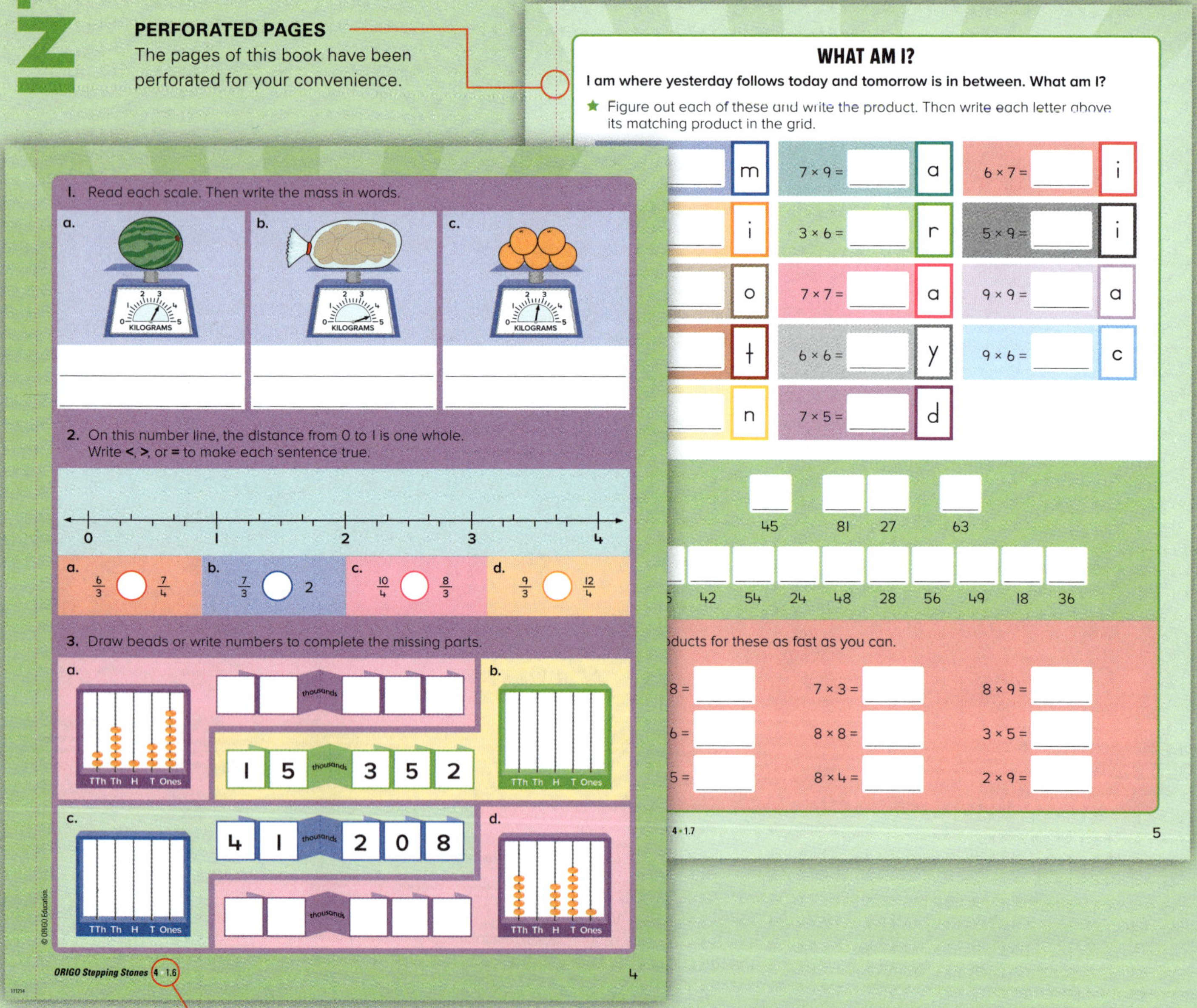


**ORIGO Stepping Stones 4 • 1.6**

Grade — Module — Lesson

INTRODUCTION

## STUDENT JOURNAL

The student journal provides a double-page spread for each lesson in the ***ORIGO Stepping Stones*** program for Grade 4. Each spread includes guided discussion of enquiry, questions based on the discussion, and a final question that puts a little twist on the content to promote higher-order thinking skills.

## NOTES FOR HOME

Each book is one component of a comprehensive teaching program. Together they are a collection of consolidation and practice pages from lessons in the ***ORIGO Stepping Stones*** program.

Class teachers will decide which pages suit individual needs. So students might not complete every page in these books. For more information about the program, visit **www.origoeducation.com/steppingstones**.

## ADDITIONAL RESOURCES – PRINT

*The Number Case* provides teachers with ready-made resources that are designed to develop students' understanding of number.

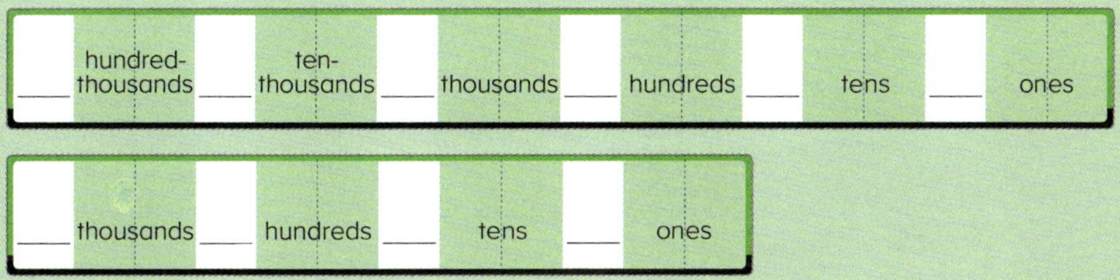


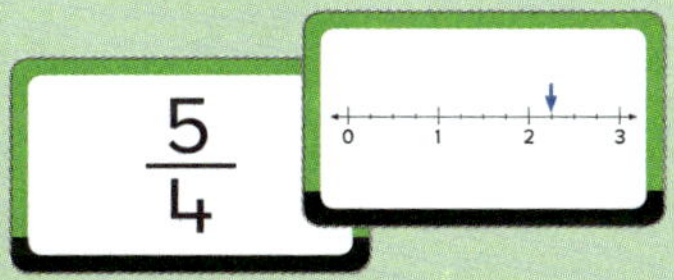


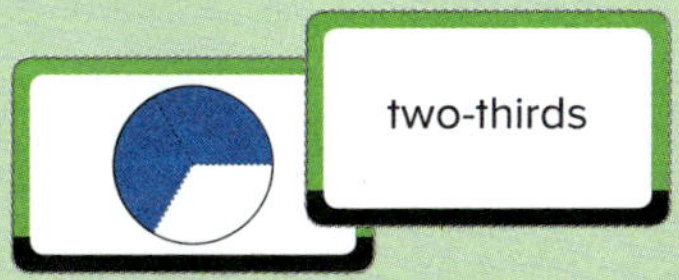


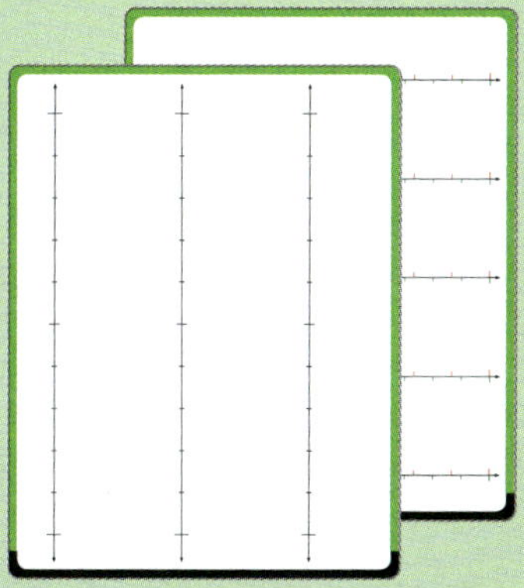

## ADDITIONAL RESOURCES (ONLINE CHANNELS)

These are some of the innovative teaching channels integrated into the teacher's online program.

**ORIGO MathEd**

Professional learning sessions

**Flare**

Interactive whiteboard tools

**Fundamentals Game Boards**

Interactive games

**1.** For each of these, use the standard addition algorithm to calculate the total cost.

**a.** $175 $263

H T O

\+

**b.** $408 $377

H T O

\+

**c.** $286 $145

H T O

\+

**2.** On each number line, the distance from 0 to 1 is one whole. Look at how each number line has been split. Write the fraction each arrow is pointing to.

**a.**

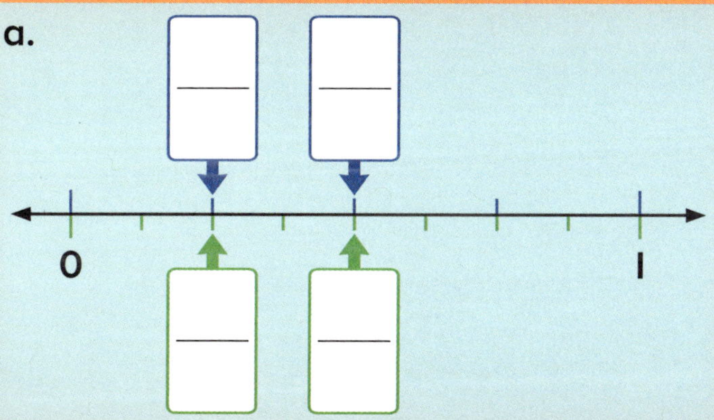


**b.**

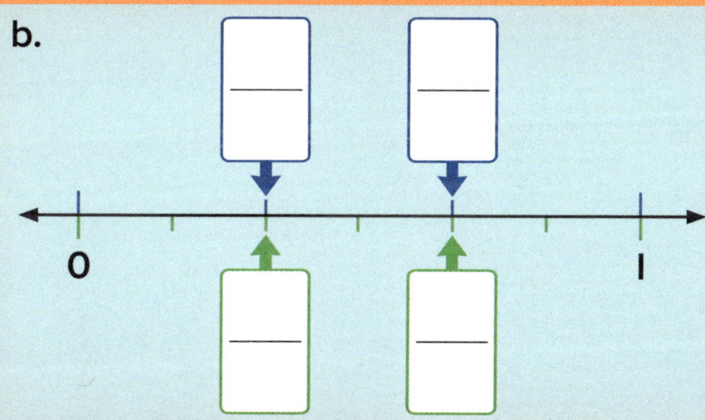


**3.** Look at the blocks. Write the matching number on the expander. Then write the number in words.

**a.**

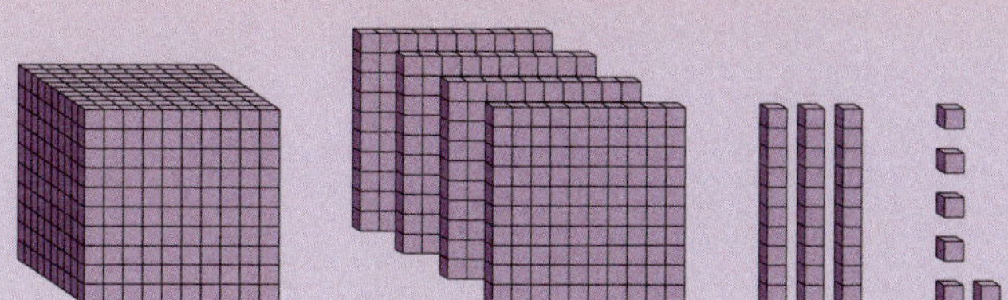

**b.**

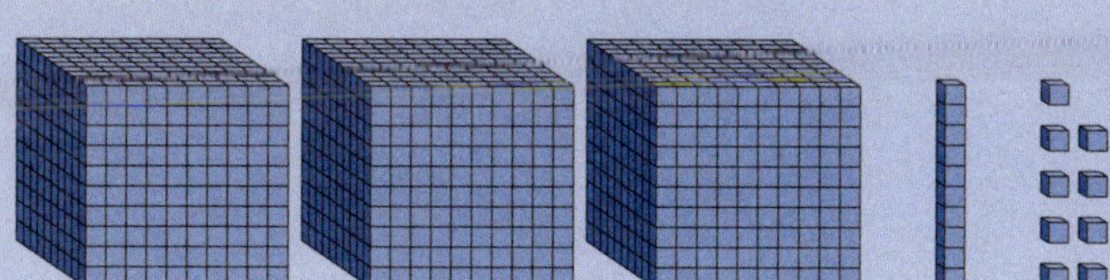

# WHERE AM I?

**When Grace was on vacation, she found a room that had no floor or ceiling and no windows or doors. What type of room did she find?**

★ Figure out each of these and write the answer. Find the answers in the grid below and cross out the letter above. Write the remaining letters at the bottom of the page.

| | | |
|---|---|---|
| 64 − 32 = ____ | 12 + 13 = ____ | 62 − 31 = ____ |
| 42 + 43 = ____ | 48 − 24 = ____ | 21 + 22 = ____ |
| 46 − 23 = ____ | 41 + 42 = ____ | 45 − 23 = ____ |
| 22 + 22 = ____ | 27 − 14 = ____ | 14 + 15 = ____ |
| 82 − 41 = ____ | 22 + 23 = ____ | 65 − 32 = ____ |
| 32 + 31 = ____ | 84 − 42 = ____ | 24 + 24 = ____ |

❋

| B | A | T | H | G | A | M | E | S |
|---|---|---|---|---|---|---|---|---|
| 23 | 30 | 42 | 24 | 45 | 31 | 14 | 41 | 63 |
| C | U | P | S | H | O | W | E | R |
| 83 | 64 | 32 | 84 | 46 | 85 | 25 | 48 | 34 |
| B | E | D | B | R | O | O | M | S |
| 22 | 29 | 33 | 44 | 13 | 18 | 66 | 28 | 43 |

☐ ☐☐☐☐☐☐☐☐

Write the remaining letters in order from the ❋ to the bottom-right corner.

1. Estimate the difference. Then use the standard subtraction algorithm to calculate the exact difference.

a.

Estimate ________

| | H | T | O |
|---|---|---|---|
| | 3 | 6 | 7 |
| − | | 4 | 5 |
| | | | |

b.

Estimate ________

| | H | T | O |
|---|---|---|---|
| | 5 | 6 | 3 |
| − | | 3 | 8 |
| | | | |

c.

Estimate ________

| | H | T | O |
|---|---|---|---|
| | 4 | 2 | 8 |
| − | | 4 | 7 |
| | | | |

d.

Estimate ________

| | H | T | O |
|---|---|---|---|
| | 7 | 3 | 2 |
| − | | 5 | 6 |
| | | | |

2. On this number line, the distance from 0 to 1 is one whole. Write the fourths in each box above the line. Then write the matching eighths below the line.

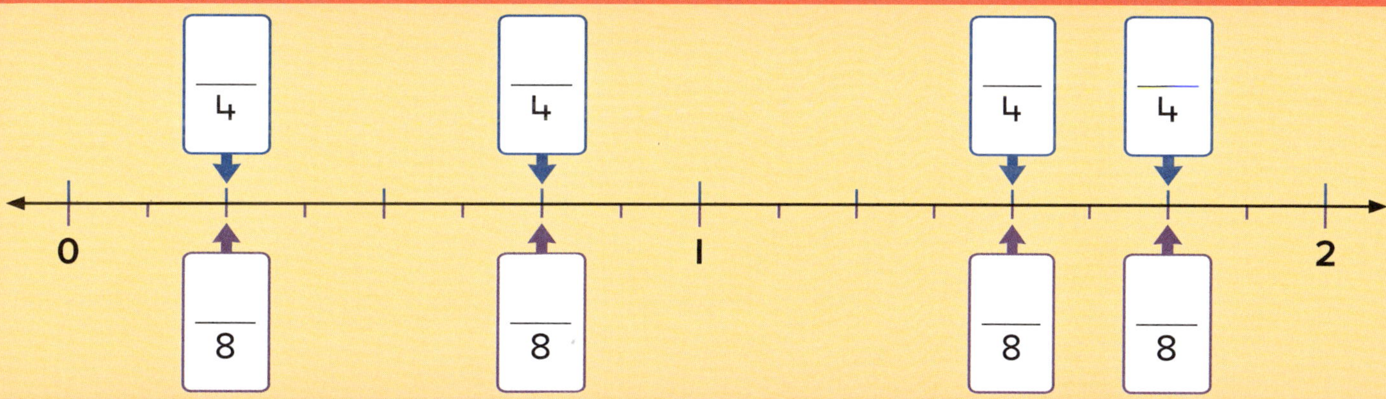


3. Write or draw the missing parts.

a.

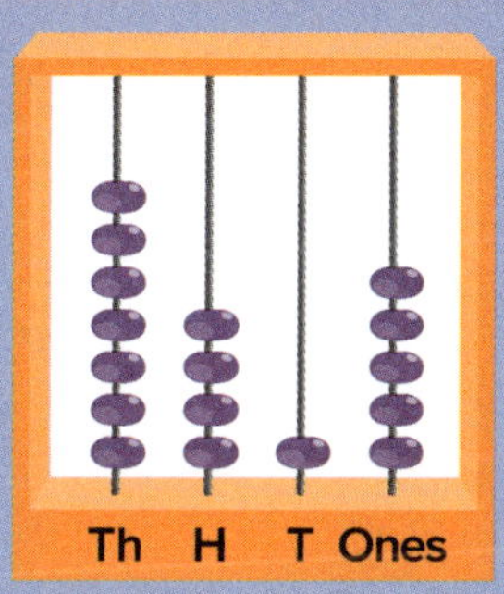


________

seven thousand ________

________

b.

2,804

________

________

**1.** Read each scale. Then write the mass in words.

**a.**

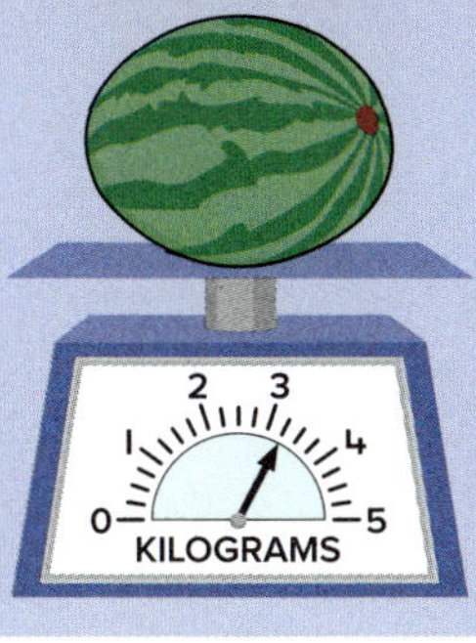


______________________

______________________

**b.**

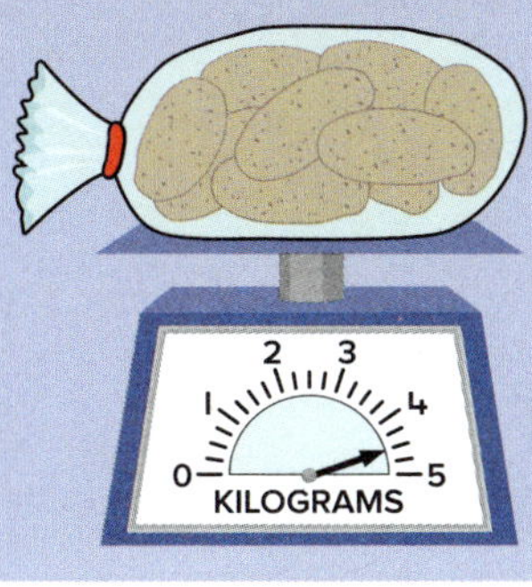


______________________

______________________

**c.**

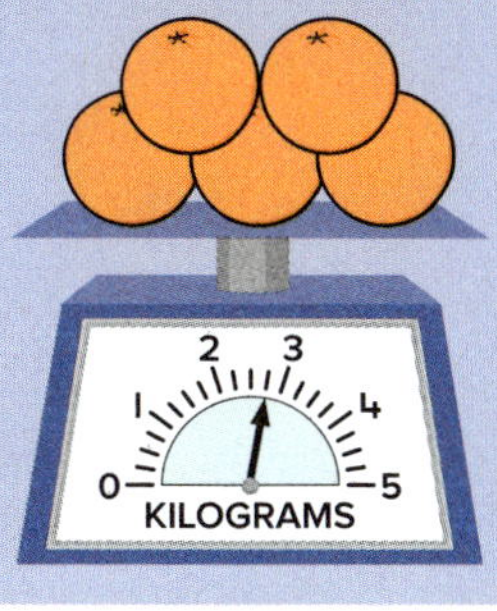


______________________

______________________

**2.** On this number line, the distance from 0 to 1 is one whole.
Write **<**, **>**, or **=** to make each sentence true.

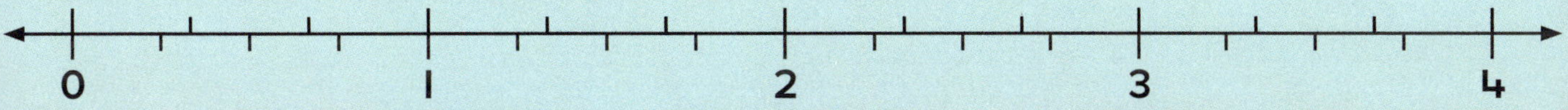


**a.** $\frac{6}{3}$ ◯ $\frac{7}{4}$

**b.** $\frac{7}{3}$ ◯ 2

**c.** $\frac{10}{4}$ ◯ $\frac{8}{3}$

**d.** $\frac{9}{3}$ ◯ $\frac{12}{4}$

**3.** Draw beads or write numbers to complete the missing parts.

**a.**

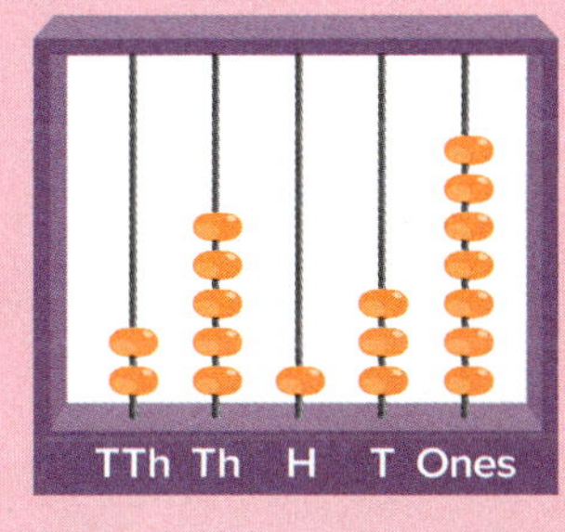


☐ ☐ thousands ☐ ☐ ☐

**b.**

1 5 thousands 3 5 2

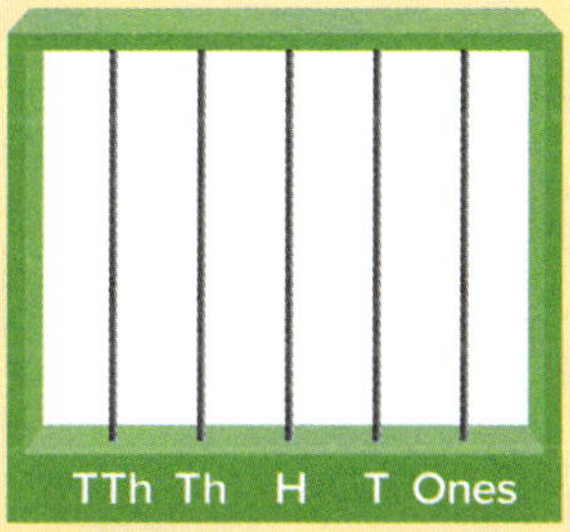


**c.**

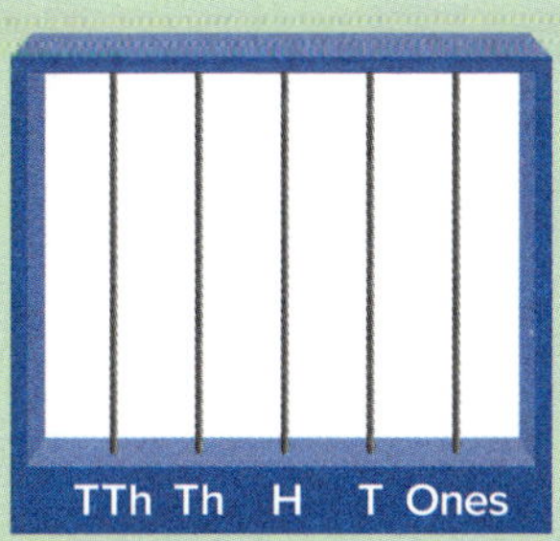


4 1 thousands 2 0 8

**d.**

☐ ☐ thousands ☐ ☐ ☐

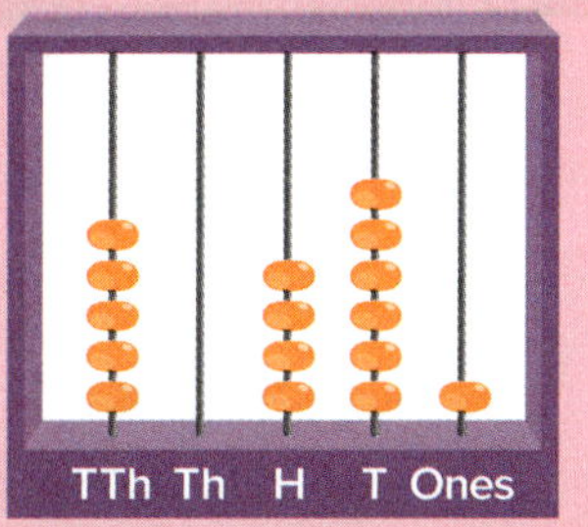

# WHAT AM I?

**I am where yesterday follows today and tomorrow is in between. What am I?**

★ Figure out each of these and write the product. Then write each letter above its matching product in the grid.

| | | |
|---|---|---|
| 9 × 3 = ____ m | 7 × 9 = ____ a | 6 × 7 = ____ i |
| 8 × 6 = ____ i | 3 × 6 = ____ r | 5 × 9 = ____ i |
| 7 × 4 = ____ o | 7 × 7 = ____ a | 9 × 9 = ____ a |
| 8 × 3 = ____ t | 6 × 6 = ____ y | 9 × 6 = ____ c |
| 8 × 7 = ____ n | 7 × 5 = ____ d | |

| ____ | | ____ | ____ | | ____ |
|---|---|---|---|---|---|
| 45 | | 81 | 27 | | 63 |

| ____ | ____ | ____ | ____ | ____ | ____ | ____ | ____ | ____ | ____ |
|---|---|---|---|---|---|---|---|---|---|
| 35 | 42 | 54 | 24 | 48 | 28 | 56 | 49 | 18 | 36 |

Write the products for these as fast as you can.

| | | |
|---|---|---|
| 5 × 8 = ____ | 7 × 3 = ____ | 8 × 9 = ____ |
| 4 × 6 = ____ | 8 × 8 = ____ | 3 × 5 = ____ |
| 6 × 5 = ____ | 8 × 4 = ____ | 2 × 9 = ____ |

**1.** Use multiplication to help you figure out the total area of each large rectangle.

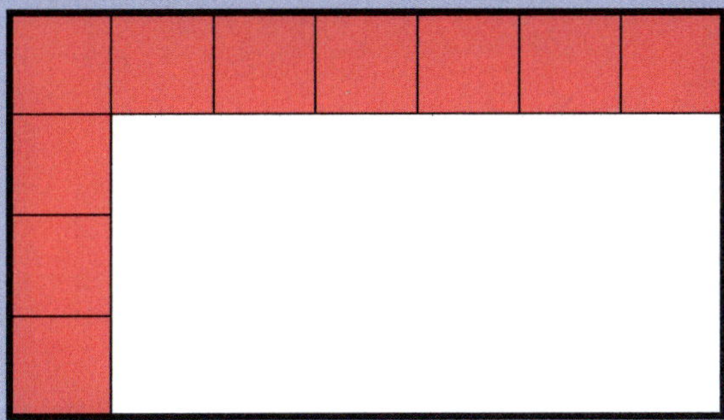

**a.** Area is ______ sq units

**b.** Area is ______ sq units

**2.** On this number line, the distance from 0 to 1 is one whole.
Complete the missing fractions below so that the order is from **least** to **greatest**.

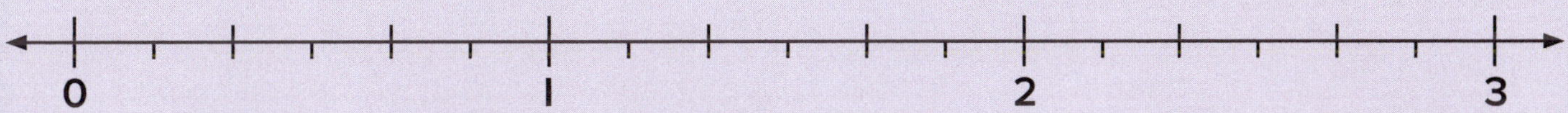


**a.** $\frac{\phantom{0}}{6}$ **b.** $\frac{1}{3}$ **c.** $\frac{\phantom{0}}{3}$ **d.** $\frac{5}{6}$ **e.** $\frac{\phantom{0}}{6}$ **f.** $\frac{7}{3}$ **g.** $\frac{\phantom{0}}{6}$

**3.** Complete the equations to describe the number shown on the expander.

**a.**

____ × 10,000 = ____________

____ × 1,000 = ____________

____ × 100 = ____________

____ × 10 = ____________

____ × 1 = ____________

**b.**

____ × 10,000 = ____________

____ × 1,000 = ____________

____ × 100 = ____________

____ × 10 = ____________

____ × 1 = ____________

**1.** Split each rectangle into two parts that are easy for you to multiply.
Then calculate the area. Show your thinking.

**a.**

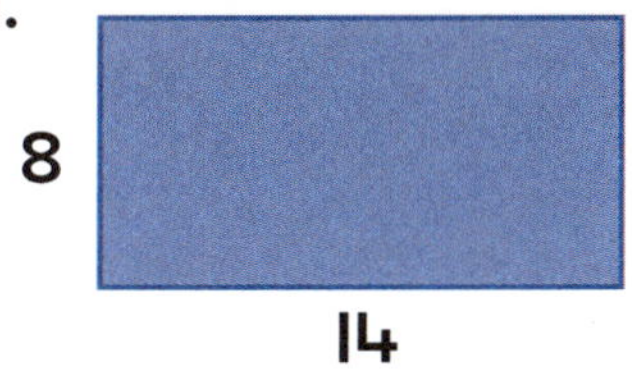


_______ sq units

**b.**

_______ sq units

**2.** Calculate the perimeter of this shape. Show your thinking.

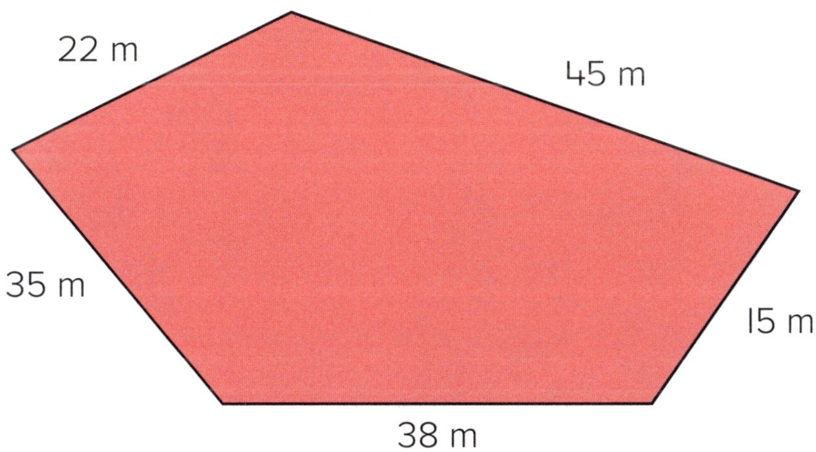


_______ m

**3.** Round each population to the nearest **hundred**.
Use the number line to help your thinking.

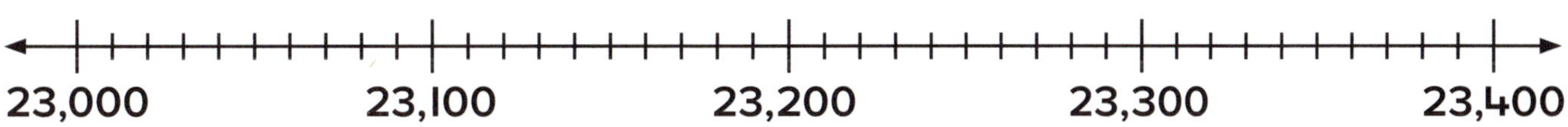


**a.** Population 23,156 _______

**b.** Population 23,308 _______

**c.** Population 23,279 _______

**d.** Population 23,120 _______

**e.** Population 23,051 _______

**f.** Population 23,015 _______

# FUN FACT

★ To discover a fun fact, figure out each of these and write the answer. Then write each letter above its matching answer at the bottom of the page. Some letters appear more than once.

| Problem | Answer | Letter | Problem | Answer | Letter |
|---|---|---|---|---|---|
| 62 + 63 = | ______ | i | 149 − 75 = | ______ | n |
| 123 − 61 = | ______ | w | 34 + 58 = | ______ | m |
| 79 + 80 = | ______ | o | 184 − 91 = | ______ | s |
| 92 − 46 = | ______ | l | 58 + 58 = | ______ | t |
| 38 + 39 = | ______ | h | 167 − 85 = | ______ | y |
| 129 − 64 = | ______ | z | 64 + 66 = | ______ | b |
| 26 + 57 = | ______ | p | 127 − 63 = | ______ | u |
| 59 − 28 = | ______ | r | 66 + 28 = | ______ | e |

___ ___   ___ ___   ___ ___ ___ ___ ___ ___ ___ ___ ___ ___

125 116   125 93   125 92 83 159 93 93 125 130 46 94

___ ___   ___ ___ ___ ___ ___ ___   ___ ___ ___ ___

116 159   93 74 94 94 65 94   62 125 116 77

___ ___ ___ ___   ___ ___ ___ ___   ___ ___ ___ ___

82 159 64 31   94 82 94 93   159 83 94 74

**1.** Draw a line to connect each time to a matching clock.

| 25 minutes past 5 | 10 past 7 | 20 minutes to 2 | 8 minutes to 3 |
|---|---|---|---|

| 7:10 | 2:52 | 5:25 | 1:40 |
|---|---|---|---|

**2.** Both of these shapes have sides that are the same length. Calculate the perimeter of each shape and write an equation to show your thinking.

**a.**

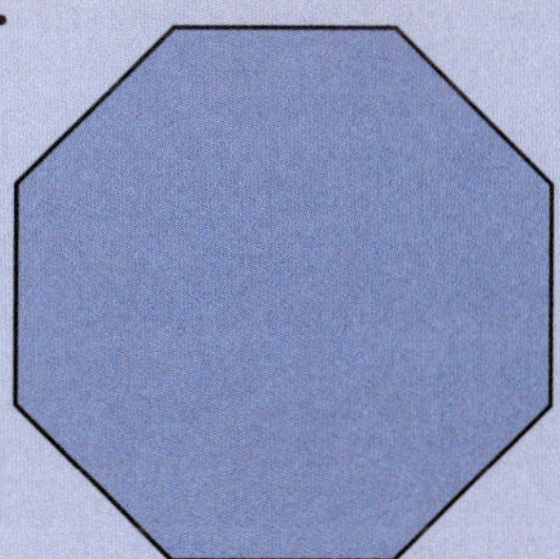

7 m

____________________ = ________ m

**b.**

15 m

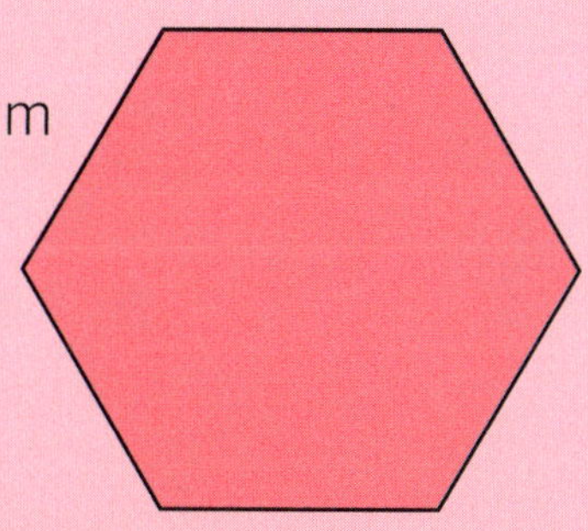

____________________ = ________ m

**3.** The table below shows the number of shapes and toothpicks for the first three pictures in this pattern. Write a word rule to show how you can figure out the number of toothpicks when you know the number of shapes.

| Picture | Number of shapes | Number of toothpicks |
|---|---|---|
| 1 | 1 | 8 |
| 2 | 2 | 16 |
| 3 | 3 | 24 |

1. Calculate the total passing yards for each of these. Show your thinking.

**PASSING YARDS**

| | |
|---|---|
| Game 1 | 346 yards |
| Game 2 | 385 yards |
| Game 3 | 292 yards |

a. Games 1 and 2

____________

b. Games 2 and 3

____________

2. Write the matching number or number words.

a. 3,952 ____________

b. two thousand four hundred sixty ____________

c. 8,714 ____________

3. Draw jumps on the number line to show how you would figure out the total. Then use words or numbers to show what you did.

a. 89 + 37 = ____________

b. 158 + 68 = ____________

# SHORE THING

**Does the ocean say anything to the beach?**

★ Figure out each of these and draw a straight line to the correct product. The line will pass through a number and a letter. Then write each letter above its matching number at the bottom of the page.

| | Products |
|---|---|
| 3 × 43 | 364 |
| 63 × 2 | 156 |
| 91 × 4 | 129 |
| 2 × 83 | 148 |
| 52 × 3 | 126 |
| 2 × 43 | 166 |
| 22 × 4 | 84 |
| 3 × 41 | 86 |
| 74 × 2 | 102 |
| 2 × 51 | 88 |
| 4 × 21 | 69 |
| 24 × 2 | 123 |
| 4 × 72 | 288 |
| 23 × 3 | 68 |
| 34 × 2 | 48 |

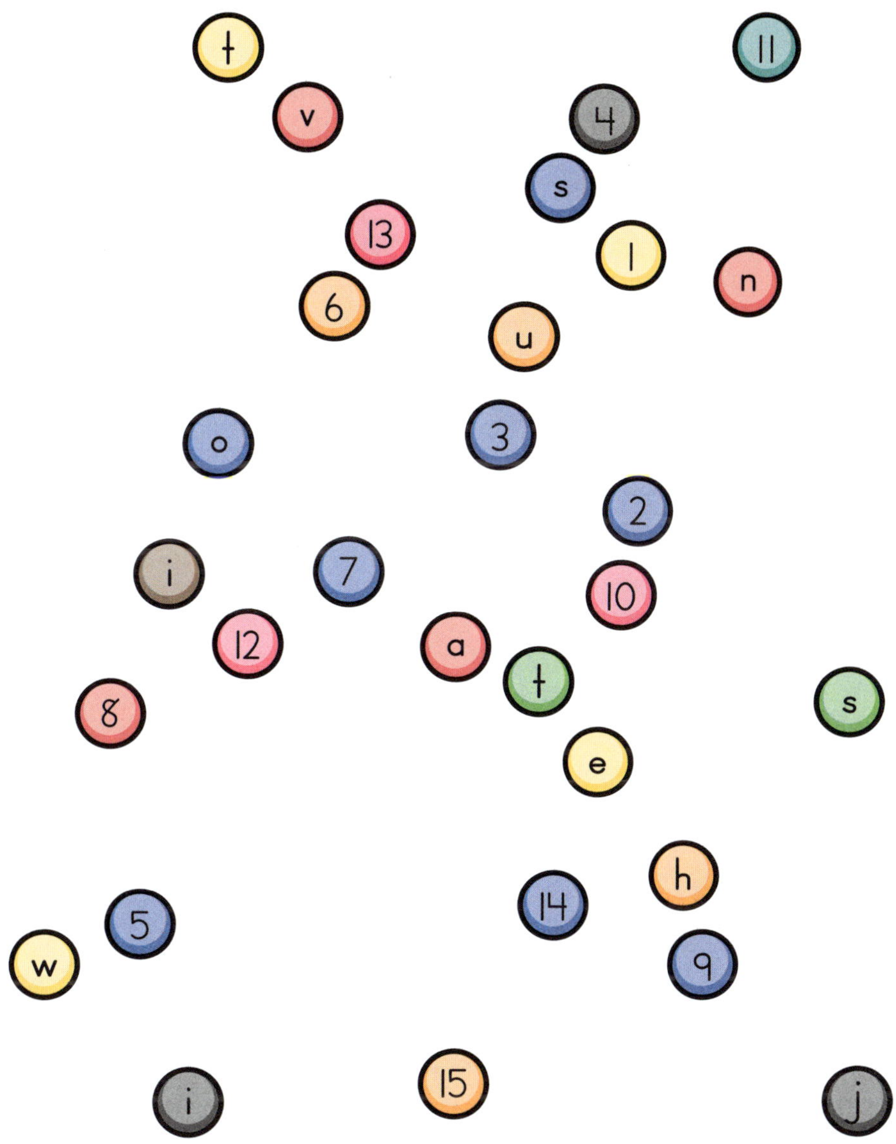


___ ___ , ___ ___ ___ ___ ___ ___

1 2 3 4 5 6 7 8

___ ___ ___ ___ ___ ___ ___ !

9 10 11 12 13 14 15

1. Each large rectangle is one whole. For each of these, write the fraction shown by the first group of shapes. Then color the same amount of space in the shapes below and write the matching fraction.

a. $\frac{18}{8}$

b.

2. Write these distances in order from **least** to **greatest**.

| | |
|---|---|
| Seattle to Las Vegas | 860 miles |
| Seattle to Dallas | 1,636 miles |
| Seattle to Atlanta | 2,187 miles |
| Seattle to New York | 2,383 miles |
| Seattle to Los Angeles | 942 miles |
| Seattle to Miami | 2,738 miles |

3. Draw jumps on the number line to show how you would figure out the difference. Then complete the equation.

a. 372 − 35 = ______

b. 245 − 89 = ______

c. 134 − 78 = ______

1. a. Color one part in each row of this fraction chart.

| 1 | | | | | | | | |
|---|---|---|---|---|---|---|---|---|
| $\frac{1}{3}$ | | | $\frac{1}{3}$ | | | $\frac{1}{3}$ | | |
| $\frac{1}{6}$ | | $\frac{1}{6}$ | $\frac{1}{6}$ | | $\frac{1}{6}$ | $\frac{1}{6}$ | | $\frac{1}{6}$ |
| $\frac{1}{9}$ | $\frac{1}{9}$ | $\frac{1}{9}$ | $\frac{1}{9}$ | $\frac{1}{9}$ | $\frac{1}{9}$ | $\frac{1}{9}$ | $\frac{1}{9}$ | $\frac{1}{9}$ |

b. Loop the fraction that is **greater** in each pair.

$\frac{1}{3}$ or $\frac{1}{6}$

$\frac{1}{6}$ or $\frac{1}{9}$

$\frac{1}{9}$ or $\frac{1}{3}$

2. Draw beads on each abacus to represent the number.

a.

32,049

b.

53,901

c.

20,375

3. Estimate each total. Then use the standard addition algorithm to calculate the exact total.

a.

Estimate __________

| | H | T | O |
|---|---|---|---|
| | 4 | 5 | 8 |
| + | | 2 | 7 |
| | | | |

b.

Estimate __________

| | H | T | O |
|---|---|---|---|
| | 3 | 5 | 6 |
| + | 2 | 3 | 8 |
| | | | |

c.

Estimate __________

| | H | T | O |
|---|---|---|---|
| | 2 | 6 | 9 |
| + | 3 | 0 | 5 |
| | | | |

d.

Estimate __________

| | H | T | O |
|---|---|---|---|
| | 5 | 2 | 7 |
| + | 2 | 5 | 4 |
| | | | |

Time Taken:

# RACE TRACK

★ For each division fact, write the multiplication fact you would use to help figure out the answer. Then write the answers. Complete the race as fast as you can. Use the classroom clock to time yourself.

$54 \div 9 =$ ____
____ × ____ = ____

$30 \div 6 =$ ____
____ × ____ = ____

$48 \div 6 =$ ____
____ × ____ = ____

$24 \div 3 =$ ____
____ × ____ = ____

$36 \div 9 =$ ____
____ × ____ = ____

$45 \div 9 =$ ____
____ × ____ = ____

$15 \div 3 =$ ____
____ × ____ = ____

$36 \div 6 =$ ____
____ × ____ = ____

$54 \div 6 =$ ____
____ × ____ = ____

$21 \div 3 =$ ____
____ × ____ = ____

$72 \div 9 =$ ____
____ × ____ = ____

$81 \div 9 =$ ____
____ × ____ = ____

$27 \div 3 =$ ____
____ × ____ = ____

$42 \div 6 =$ ____
____ × ____ = ____

$24 \div 6 =$ ____
____ × ____ = ____

$18 \div 3 =$ ____
____ × ____ = ____

$63 \div 9 =$ ____
____ × ____ = ____

1. Use the standard subtraction algorithm to calculate how much the price has dropped. Remember to estimate the answer to check that it makes sense.

a. Was $265 Now $182

| | H | T | O |
|---|---|---|---|
| | | | |
| − | | | |
| | | | |

b. Was $405 Now $268

| | H | T | O |
|---|---|---|---|
| | | | |
| − | | | |
| | | | |

c. Was $748 Now $604

| | H | T | O |
|---|---|---|---|
| | | | |
| − | | | |
| | | | |

2. Complete the missing parts.

a. 26,308 ☐ ☐ thousands ☐ hundreds ☐ tens ☐ ones

_______________________________________________

b. ____________ ☐ ☐ thousands ☐ hundreds ☐ tens ☐ ones

**fourteen thousand one hundred sixty-nine**

3. Estimate each total. Then use the standard addition algorithm to calculate the exact total.

a. Estimate ____________

| | | | |
|---|---|---|---|
| | 3 | 6 | 4 |
| + | 2 | 5 | 3 |
| | | | |

b. Estimate ____________

| | | | |
|---|---|---|---|
| | 4 | 7 | 3 |
| + | 1 | 4 | 6 |
| | | | |

c. Estimate ____________

| | | | |
|---|---|---|---|
| | 2 | 8 | 3 |
| + | | 6 | 5 |
| | | | |

d. Estimate ____________

| | | | |
|---|---|---|---|
| | 3 | 5 | 8 |
| + | 2 | 7 | 9 |
| | | | |

1. Complete each table to show the equivalent fractions and whole numbers. Use multiplication patterns to help you.

a.

| Fraction | $\frac{2}{2}$ | $\frac{}{2}$ | $\frac{}{2}$ | $\frac{8}{2}$ | $\frac{10}{2}$ | $\frac{}{2}$ | $\frac{}{2}$ |
|---|---|---|---|---|---|---|---|
| Whole number | 1 | 2 | 3 | | | 8 | 10 |

b.

| Fraction | $\frac{5}{5}$ | $\frac{}{5}$ | $\frac{}{5}$ | $\frac{20}{5}$ | $\frac{}{5}$ | $\frac{40}{5}$ | $\frac{}{5}$ |
|---|---|---|---|---|---|---|---|
| Whole number | | 2 | 3 | | 6 | | 10 |

2. Write these stadium capacities in order from **greatest** to **least**.

| Stadium | Seating Capacity |
|---|---|
| Citrus Bowl, FL | 65,438 |
| Georgia Dome, GA | 71,250 |
| Gillette Stadium, MA | 68,756 |
| Lambeau Field, WI | 72,928 |
| Memorial Stadium, NE | 81,067 |
| Spartan Stadium, MI | 75,025 |

_______ → _______ → _______
_______ ← _______ ← _______

3. Estimate each total. Then use the standard addition algorithm to calculate the exact total.

a. Estimate _______

```
  3 6 0 8
+ 4 2 3 6
```

b. Estimate _______

```
  3 5 6 8 2
+   4 1 9 5
```

c. Estimate _______

```
  4 0 9 6 2
+ 2 1 4 3 7
```

# DIRECTIONS

**If Carlos stood facing south with his back to the north, what would be on his left hand?**

★ Figure out each of these. Then write each letter above its matching answer at the bottom of the page. Some letters appear more than once.

| | | | |
|---|---|---|---|
| 220 + 350 = ______ | o | 670 − 350 = ______ | e |
| 760 − 240 = ______ | n | 260 + 520 = ______ | u |
| 510 + 330 = ______ | s | 790 − 450 = ______ | a |
| 560 − 340 = ______ | h | 210 + 460 = ______ | d |
| 450 + 310 = ______ | i | 970 − 420 = ______ | b |
| 580 − 170 = ______ | g | 240 + 530 = ______ | r |
| 350 + 140 = ______ | m | 650 − 410 = ______ | f |
| 780 − 430 = ______ | t | | |

____ ____ ____ ____    ____ ____ ____ ____ ____ ____ ____

240 570 780 770    240 760 520 410 320 770 840

____ ____ ____    ____    ____ ____ ____ ____ ____

340 520 670    340    350 220 780 490 550

**1.** Draw a simple picture to match each story. Then label the dimensions on your picture and write the answer.

**a.** Dad is building a fence around our rectangular vegetable garden that measures 16 ft by 6 ft. What is the perimeter of the garden?

________ ft

**b.** A square picture frame has a perimeter of 144 cm. What is the length of each side?

________ cm

**2.** Round each price to the nearest **hundred** dollars.

| | | |
|---|---|---|
| **a.** $16,499<br>$________ | **b.** $22,320<br>$________ | **c.** $17,455<br>$________ |
| **d.** $31,095<br>$________ | **e.** $19,990<br>$________ | **f.** $17,040<br>$________ |

**3.** Show the time on the analog and digital clocks.

**a.** 20 minutes to 4

**b.** 15 minutes to 11

**c.** 8 minutes to 5

**d.** 25 minutes to 12

1. Calculate the part in the parentheses and write the new problem. Then write the total.

a. 25 + (6 × 5)

= 25 + 30

= 55

b. (30 + 6) ÷ 4

= ______ ÷ ______

= ______

c. (7 − 4) × 8

= ______ × ______

= ______

d. 30 − (30 ÷ 10)

= ______ − ______

= ______

e. 18 + (16 ÷ 2)

= ______ + ______

= ______

f. 200 + (3 × 7)

= ______ + ______

= ______

2. **Add** the prices in each pair. Write equations to record the steps you use.

a. $176 $253

Total $ ______

b. $328 $149

Total $ ______

c. $295 $357

Total $ ______

3. Write numbers to complete each of these.

a. 24 is a multiple of ______ / ______ is a factor of 24

b. 42 is a multiple of ______ / ______ is a factor of 42

c. ______ is a multiple of 5 / 5 is a factor of ______

d. ______ is a multiple of 3 / 3 is a factor of ______

e. 30 is a multiple of ______ / ______ is a factor of 30

f. ______ is a multiple of 9 / 9 is a factor of ______

# TINY FEATHERS

**Which bird is so tiny that one of its enemies is a praying mantis?**

★ Figure out each of these and write the answer. Then write each letter above its matching answer at the bottom of the page.

| | | |
|---|---|---|
| $(5 + 14) \times 8 =$ | ______ | m |
| $5 \times (8 + 45) =$ | ______ | h |
| $(25 + 7) \times 4 =$ | ______ | e |
| $3 \times (15 + 7) =$ | ______ | t |
| $(45 + 8) \times 4 =$ | ______ | u |
| $8 \times (35 + 7) =$ | ______ | h |
| $(4 + 35) \times 6 =$ | ______ | g |
| $4 \times (3 + 15) =$ | ______ | b |
| $(75 + 6) \times 3 =$ | ______ | n |
| $5 \times (6 + 35) =$ | ______ | i |
| $(25 + 9) \times 8 =$ | ______ | i |
| $6 \times (15 + 8) =$ | ______ | r |
| $(45 + 8) \times 8 =$ | ______ | d |
| $4 \times (9 + 25) =$ | ______ | m |

Working Space

| ___ | ___ | ___ |
|---|---|---|
| 66 | 265 | 128 |

| ___ | ___ | ___ | ___ | ___ | ___ | ___ | ___ | ___ | ___ | ___ |
|---|---|---|---|---|---|---|---|---|---|---|
| 336 | 212 | 136 | 152 | 272 | 243 | 234 | 72 | 205 | 138 | 424 |

1. For each abacus, draw beads to show the number. Then use a different color to draw one more bead **in the place shown** and write the new number.

a. 3,506

the tens place

__________

b. 2,851

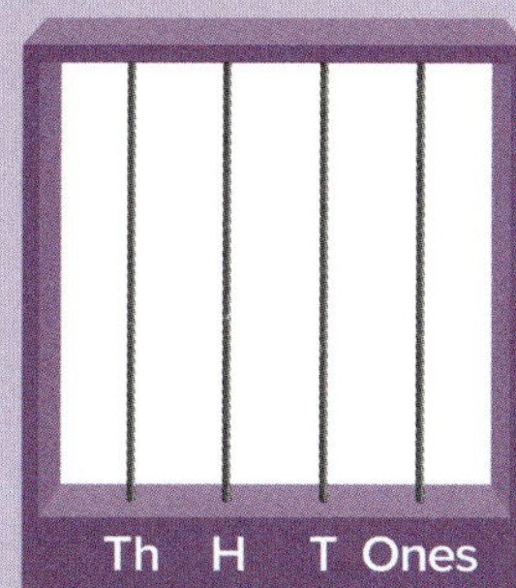


the hundreds place

__________

c. 6,075

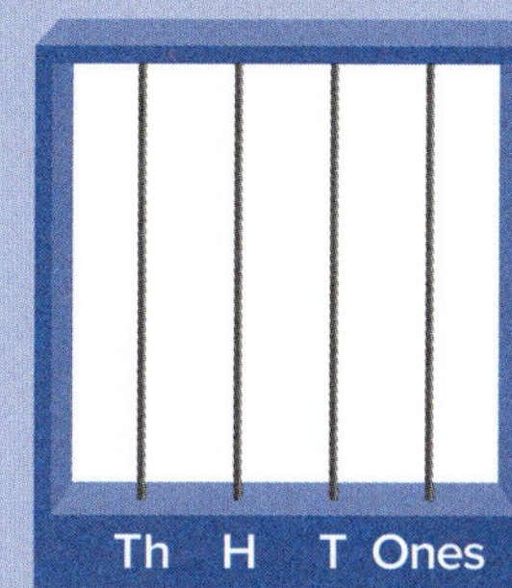


the thousands place

__________

2. Figure out the difference in value. Record the steps you use.

a. $463 $148

$ ______

b. $217 $185

$ ______

c. $351 $265

$ ______

3. Write all the factors of each number.

a. 18 is the same as

___ × ___

___ × ___

___ × ___

b. 50 is the same as

___ × ___

___ × ___

___ × ___

c. 32 is the same as

___ × ___

___ × ___

___ × ___

d. 40 is the same as

___ × ___

___ × ___

___ × ___

___ × ___

1. Write **<** or **>** to make each sentence true.

a. $\frac{15}{5}$ ◯ $\frac{20}{5}$

b. $\frac{18}{3}$ ◯ $\frac{9}{3}$

c. $\frac{9}{4}$ ◯ $\frac{11}{4}$

d. $\frac{8}{6}$ ◯ $\frac{5}{6}$

e. $\frac{16}{8}$ ◯ $\frac{12}{8}$

f. $\frac{7}{2}$ ◯ $\frac{12}{2}$

2. Estimate each total. Then use the standard addition algorithm to calculate the exact total.

a. Estimate ______

```
   2 6 1 8
 + 1 2 6 5
 ---------
```

b. Estimate ______

```
   4 6 5 7
 + 3 7 0 2
 ---------
```

c. Estimate ______

```
   3 2 8 9
 + 2 0 1 5
 ---------
```

3. Color all the composite numbers. Then color a matching array to prove that each number is composite.

18 11 21 15 17

# CROSS NUMBER

★ Write the total for each of these in the puzzle grid below.

| Across | Down |
|---|---|
| **a.** 200 + 15 + 30 | **a.** 200 + 56 + 30 |
| **c.** 200 + 53 + 40 | **b.** 500 + 28 + 30 |
| **e.** 500 + 57 + 30 | **c.** 200 + 44 + 30 |
| **f.** 600 + 48 + 30 | **d.** 300 + 34 + 20 |
| **h.** 400 + 64 + 20 | **g.** 700 + 28 + 50 |
| **j.** 300 + 55 + 30 | **i.** 800 + 66 + 10 |
| **l.** 100 + 34 + 30 | **j.** 300 + 71 + 20 |
| **n.** 700 + 25 + 40 | **k.** 500 + 56 + 20 |
| **o.** 100 + 36 + 40 | **l.** 100 + 22 + 30 |
| **p.** 200 + 45 + 10 | **m.** 400 + 65 + 20 |

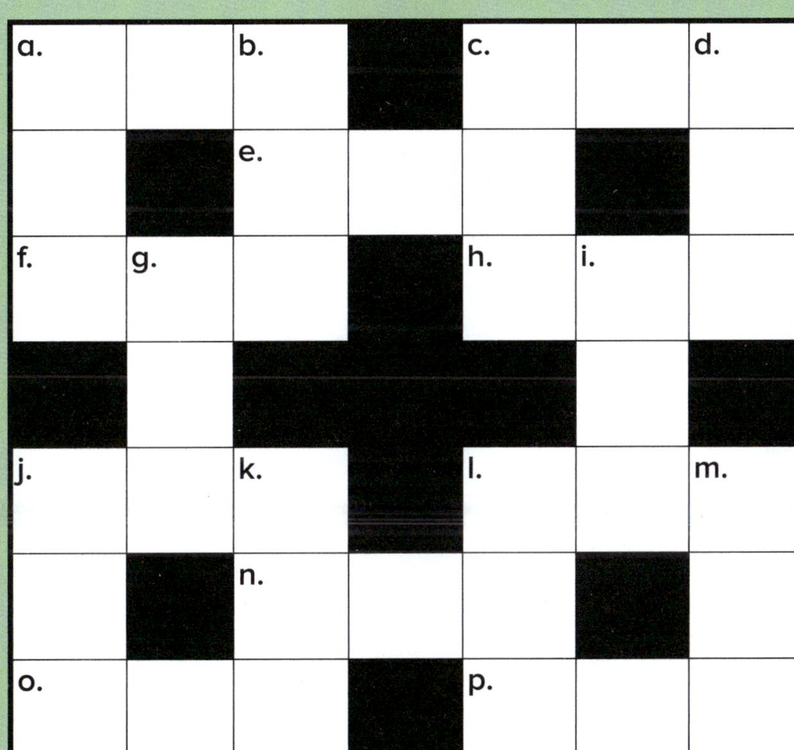

1. Complete the equations to describe the number shown on the expander.

a. 3 8 2 5 4

____ × 10,000 = ____________

____ × 1,000 = ____________

____ × 100 = ____________

____ × 10 = ____________

____ × 1 = ____________

b. 4 1 3 0 7

____ × 10,000 = ____________

____ × 1,000 = ____________

____ × 100 = ____________

____ × 10 = ____________

____ × 1 = ____________

2. Estimate each total. Then use the standard addition algorithm to calculate the exact total.

a. Estimate ____________

|   |   | 4 | 3 | 6 | 0 |
|---|---|---|---|---|---|
|   |   |   | 8 | 0 | 4 |
| + |   |   | 2 | 7 | 3 |

b. Estimate ____________

|   | 3 | 6 | 0 | 2 | 0 |
|---|---|---|---|---|---|
|   |   | 2 | 6 | 5 | 4 |
| + |   | 1 | 1 | 6 | 8 |

c. Estimate ____________

|   | 5 | 2 | 7 | 8 | 9 |
|---|---|---|---|---|---|
|   | 1 | 3 | 5 | 4 | 0 |
| + |   | 3 | 4 | 2 | 0 |

3. For each prism, write the dimensions in the order that you would multiply. Then write the total number of cubes.

a.

____ × ____ × ____ = ____________

b.

____ × ____ × ____ = ____________

**1.** Round each car price to the nearest **ten**, **hundred**, and **thousand** dollars.

| | Car Prices | | | | |
|---|---|---|---|---|---|
| | $15,959 | $21,089 | $29,356 | $24,612 | $46,111 |
| Nearest ten | | | | | |
| Nearest hundred | | | | | |
| Nearest thousand | | | | | |

**2.** Figure out the total number of minutes. Then write the total another way.

**a.**

| | |
|---|---|
| Washed Mom's car | 37 minutes |
| Washed Dad's car | 45 minutes |

_______ minutes

____ hour _______ minutes

**b.**

| | |
|---|---|
| Watched TV | 58 minutes |
| Did homework | 49 minutes |

_______ minutes

____ hour _______ minutes

**3.** Each large shape is one whole. Shade parts to show each fraction.

**a.** $\frac{5}{8}$

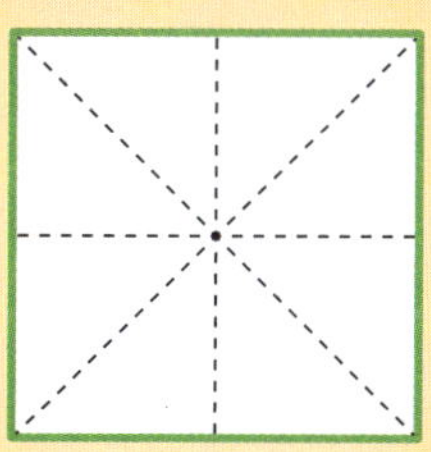

**b.** $\frac{7}{10}$

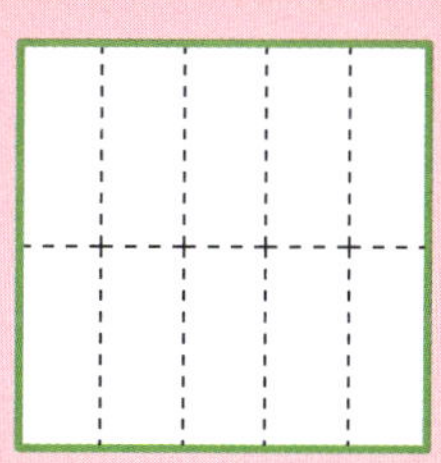

**c.** $\frac{3}{5}$

**d.** $\frac{1}{4}$

**e.** $\frac{2}{10}$

**f.** $\frac{3}{4}$

**g.** $\frac{7}{8}$

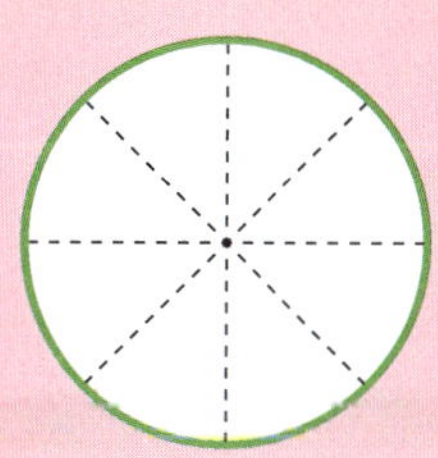

**h.** $\frac{2}{5}$

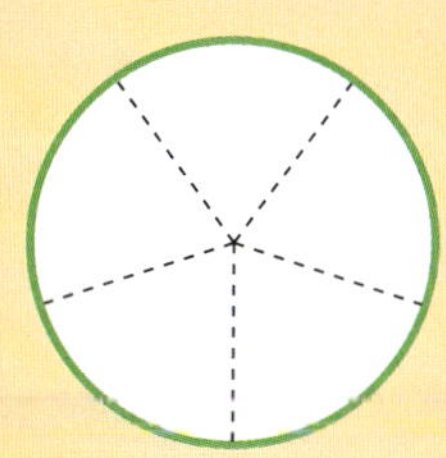

**i.** $\frac{3}{8}$

**j.** $\frac{4}{5}$

**k.** $\frac{6}{10}$

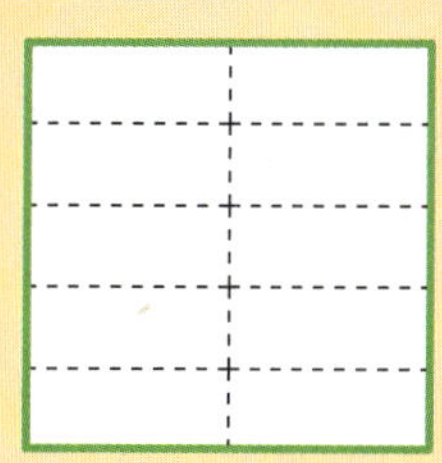

**l.** $\frac{1}{2}$

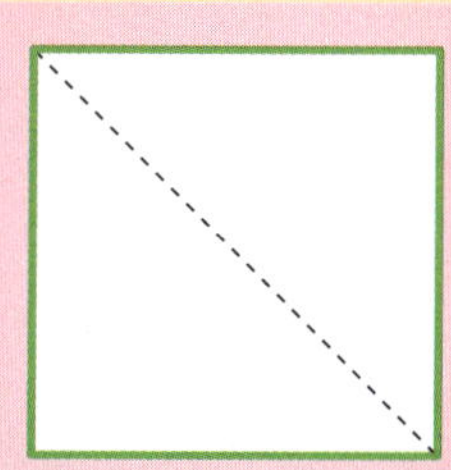

# BEDTIME

**What do you take off just before getting into bed?**

★ Figure out each of these and write the product. Then write each letter above its matching product at the bottom of the page.

| | | |
|---|---|---|
| | | 8 × 4 × 2 = ____ o |
| 5 × 3 × 7 = ____ f | 6 × 2 × 8 = ____ e | 4 × 5 × 6 = ____ u |
| 7 × 2 × 9 = ____ e | 2 × 5 × 8 = ____ o | 5 × 6 × 2 = ____ f |
| 9 × 5 × 3 = ____ y | 7 × 5 × 2 = ____ h | 4 × 7 × 3 = ____ r |
| 7 × 6 × 4 = ____ r | 4 × 7 × 5 = ____ o | 8 × 9 × 4 = ____ t |
| 7 × 4 × 8 = ____ t | 3 × 6 × 4 = ____ o | 9 × 3 × 4 = ____ f |
| 9 × 3 × 2 = ____ f | 8 × 5 × 4 = ____ l | 2 × 0 × 7 = ____ e |

## Working Space

____ ____ ____ ____   ____ ____ ____ ____   ____ ____ ____

135 72 120 168   54 0 126 224   140 105 60

____ ____ ____   ____ ____ ____ ____ ____

288 70 96   108 160 80 64 84

**1.** Look at this pattern.

Picture 1

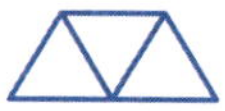
Picture 2

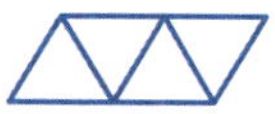
Picture 3

Picture 4

**a.** Draw the next two pictures in the pattern.

Picture 5

Picture 6

**b.** Complete this table to match.

| **Picture** | 1 | 2 | | | | |
|---|---|---|---|---|---|---|
| **Number of triangles** | 2 | | | | | |

**2.** These start and finish times are in the morning on the same day.
Write the number of seconds that have passed.

**a.**

______ seconds

**b.**

______ seconds

**3.** Use the number line to help you write equivalent fractions.

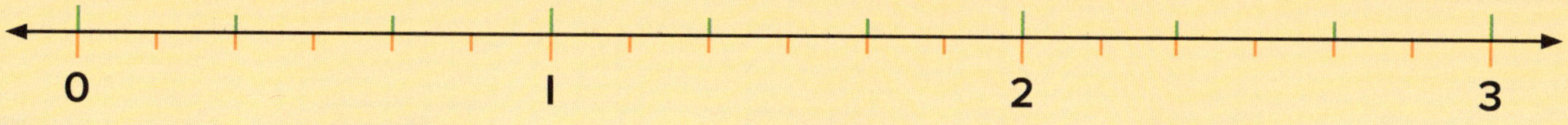


**a.** $\frac{1}{3}$ = ____

**b.** ____ = $\frac{10}{6}$

**c.** $\frac{7}{3}$ = ____

**d.** ____ = $\frac{16}{6}$

**1.** Complete the missing parts.

**a.**

**80,716**

| | | thousands | | hundreds | | tens | | ones |
|---|---|---|---|---|---|---|---|---|
| | | | | | | | | |

______________________________

**b.**

______________

| | | thousands | | hundreds | | tens | | ones |
|---|---|---|---|---|---|---|---|---|
| | | | | | | | | |

**twenty-nine thousand five hundred three**

**2.** Double one number and halve the other. Then write the products.

**a.** $18 \times 6 =$ ______
is the same as
______ × ______ = ______

**b.** $6 \times 15 =$ ______
is the same as
______ × ______ = ______

**c.** $4 \times 35 =$ ______
is the same as
______ × ______ = ______

**d.** $45 \times 8 =$ ______
is the same as
______ × ______ = ______

**e.** $5 \times 24 =$ ______
is the same as
______ × ______ = ______

**f.** $3 \times 28 =$ ______
is the same as
______ × ______ = ______

**3.** Estimate the difference. Then use the standard subtraction algorithm to calculate the exact difference.

**a.** Estimate ______

| | H | T | O |
|---|---|---|---|
| | 6 | 3 | 7 |
| − | 1 | 8 | 5 |
| | | | |

**b.** Estimate ______

| | H | T | O |
|---|---|---|---|
| | 3 | 8 | 2 |
| − | 1 | 6 | 9 |
| | | | |

**c.** Estimate ______

| | H | T | O |
|---|---|---|---|
| | 7 | 9 | 3 |
| − | 5 | 5 | 7 |
| | | | |

**d.** Estimate ______

| | H | T | O |
|---|---|---|---|
| | 8 | 3 | 6 |
| − | 4 | 7 | 1 |
| | | | |

# MONSTER MASH

**I have 4 noses, 6 ears, and 5 mouths. What am I?**

★ Figure out each of these and write the answer. Find the answer in the grid below and cross out the letter above. Then write the remaining letters at the bottom of the page.

| | |
|---|---|
| 6 × (150 + 8) = ______ | 2 × (6 + 450) = ______ |
| (340 + 7) × 2 = ______ | (3 + 260) × 5 = ______ |
| 5 × (370 + 2) = ______ | 6 × (9 + 180) = ______ |
| (8 + 360) × 2 = ______ | (410 + 7) × 4 = ______ |
| 4 × (560 + 3) = ______ | 6 × (220 + 5) = ______ |
| (7 + 350) × 4 = ______ | (320 + 9) × 3 = ______ |
| (230 + 8) × 4 = ______ | 4 × (240 + 6) = ______ |
| 3 × (350 + 5) = ______ | (440 + 5) × 3 = ______ |
| (520 + 8) × 4 = ______ | 4 × (7 + 450) = ______ |

Write the remaining letters in order from the ❄ to the bottom-right corner.

❄

| V | T | E | R | R | I | B | L | E |
|---|---|---|---|---|---|---|---|---|
| 996 | 1,315 | 1,896 | 1,065 | 1,370 | 1,828 | 1,350 | 1,335 | 912 |
| F | U | N | N | Y | S | C | A | T |
| 1,134 | 1,860 | 948 | 984 | 738 | 1,085 | 962 | 1,668 | 2,112 |
| H | A | I | R | Y | U | G | L | Y |
| 694 | 2,224 | 987 | 1,081 | 2,252 | 952 | 1,428 | 736 | 696 |

| | | | |
|---|---|---|---|
| | | | |

| | | | | |
|---|---|---|---|---|
| | | | | |

**1.** Write **<** or **>** to make each sentence true.

| | | | |
|---|---|---|---|
| **a.** | 23,715 | ◯ | 23,751 |
| **b.** | 16,059 | ◯ | 15,999 |
| **c.** | 75,210 | ◯ | 57,899 |
| **d.** | 10,502 | ◯ | 10,512 |
| **e.** | 35,650 | ◯ | 35,750 |
| **f.** | 19,007 | ◯ | 19,700 |
| **g.** | 21,650 | ◯ | 22,950 |
| **h.** | 49,725 | ◯ | 49,752 |
| **i.** | 81,040 | ◯ | 79,909 |

**2.** Complete these factor trees.

**a.**

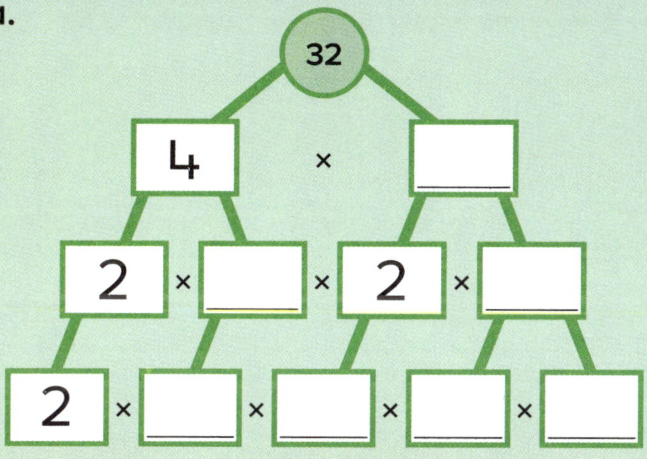


**b.**

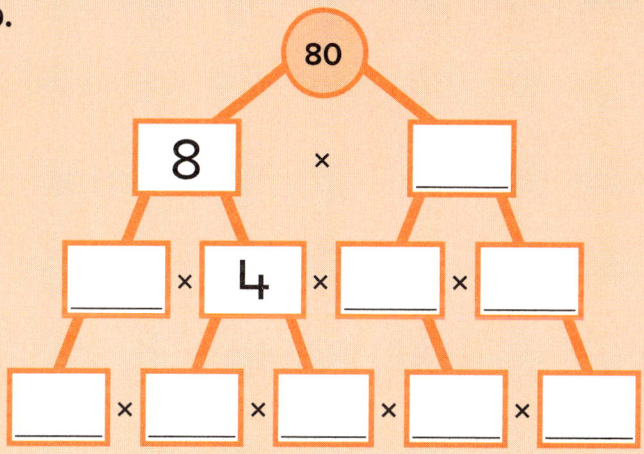


**c.**

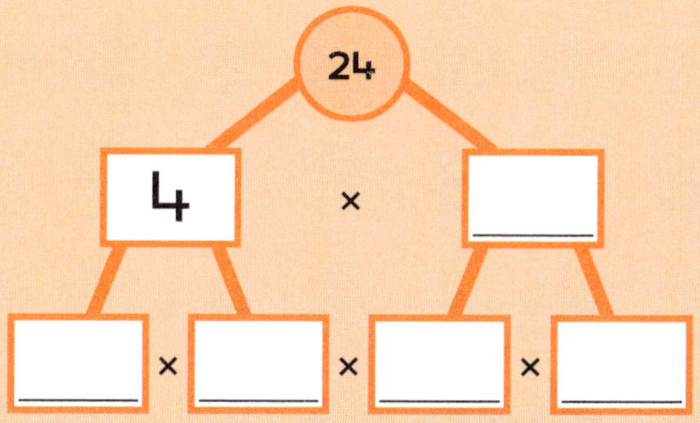


**d.**

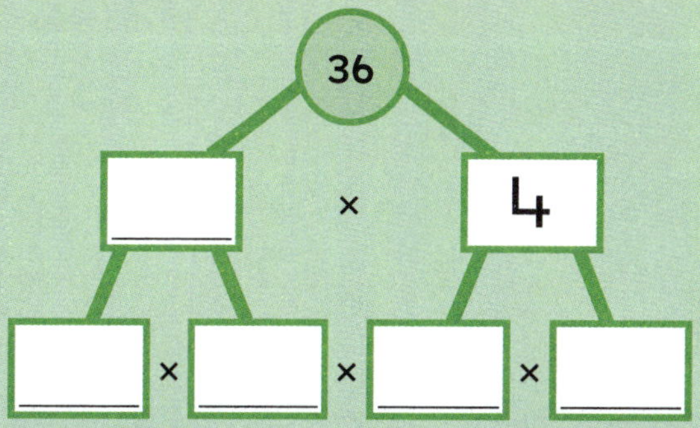


**3.** Estimate the difference. Then use the standard subtraction algorithm to calculate the exact difference.

**a.** Estimate ______

| | H | T | O |
|---|---|---|---|
| | 5 | 3 | 6 |
| − | 2 | 7 | 7 |
| | | | |

**b.** Estimate ______

| | H | T | O |
|---|---|---|---|
| | 3 | 5 | 1 |
| − | 1 | 8 | 4 |
| | | | |

**c.** Estimate ______

| | H | T | O |
|---|---|---|---|
| | 4 | 6 | 7 |
| − | 1 | 9 | 8 |
| | | | |

**d.** Estimate ______

| | H | T | O |
|---|---|---|---|
| | 5 | 1 | 4 |
| − | 2 | 1 | 6 |
| | | | |

1. Draw jumps on the number line to show how you would figure out the total. Write the total. Then use words or numbers to show what you did.

a.

146 + 37 = ____

b.

28 + 259 = ____

2. Draw a factor tree for each number.

a.

b.

3. Estimate the difference. Then use the standard subtraction algorithm to calculate the exact difference.

a. Estimate ____

| | Th | H | T | O |
|---|---|---|---|---|
| | 4 | 8 | 5 | 9 |
| – | 1 | 6 | 4 | 3 |
| | | | | |

b. Estimate ____

| | Th | H | T | O |
|---|---|---|---|---|
| | 5 | 7 | 2 | 6 |
| – | | 5 | 8 | 3 |
| | | | | |

c. Estimate ____

| | TTh | Th | H | T | O |
|---|---|---|---|---|---|
| | 3 | 5 | 8 | 6 | 9 |
| – | | 8 | 9 | 4 | 6 |
| | | | | | |

# BAREBACK

**What horse never wears a saddle?**

★ Figure out each of these and write the difference. Find each difference in the grid below and cross out the letter above. Then write the remaining letters at the bottom of the page.

| | | |
|---|---|---|
| 360 − 179 = ______ | 420 − 159 = ______ | 510 − 389 = ______ |
| 620 − 279 = ______ | 750 − 239 = ______ | 380 − 299 = ______ |
| 520 − 369 = ______ | 680 − 389 = ______ | 720 − 469 = ______ |
| 360 − 289 = ______ | 530 − 179 = ______ | 670 − 499 = ______ |
| 730 − 589 = ______ | 440 − 169 = ______ | 460 − 369 = ______ |
| 560 − 499 = ______ | 630 − 149 = ______ | 760 − 179 = ______ |

✻

| A | S | T | A | L | L | I | O | N |
|---|---|---|---|---|---|---|---|---|
| 51 | 131 | 181 | 71 | 481 | 81 | 61 | 351 | 581 |
| M | U | L | E | F | O | A | L | H |
| 171 | 121 | 291 | 191 | 141 | 261 | 281 | 151 | 321 |
| P | O | N | Y | R | I | S | E | N |
| 341 | 161 | 91 | 511 | 211 | 271 | 201 | 521 | 251 |

**1.** Complete each of these. Use the number line to show your thinking.

**a.** 384 − 59 = ______

**b.** 148 − 85 = ______

**c.** 263 − 28 = ______

**2.** Draw lines to connect three digits that are easy to multiply. Use each digit only once. Then write a number sentence to show how you would figure out the product.

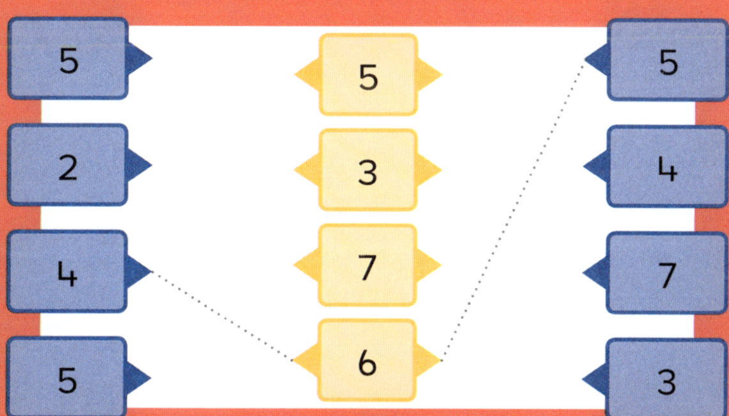


**a.** 4 × 5 × 6 = ______

**b.** ___ × ___ × ___ = ______

**c.** ___ × ___ × ___ = ______

**d.** ___ × ___ × ___ = ______

**3.** Estimate the difference. Then use the standard subtraction algorithm to calculate the exact difference.

**a.** Estimate ______

| | Th | H | T | O |
|---|---|---|---|---|
| | 7 | 5 | 0 | 6 |
| − | 1 | 3 | 8 | 4 |
| | | | | |

**b.** Estimate ______

| | TTh | Th | H | T | O |
|---|---|---|---|---|---|
| | 2 | 6 | 0 | 7 | 9 |
| − | | 8 | 5 | 1 | 6 |
| | | | | | |

**c.** Estimate ______

| | TTh | Th | H | T | O |
|---|---|---|---|---|---|
| | 1 | 0 | 3 | 8 | 5 |
| − | | 6 | 7 | 2 | 1 |
| | | | | | |

1. Solve these word problems. Show your thinking.

a. There were 3,605 senior tickets, 14,034 adult tickets, and 23,762 student tickets sold at the gate. How many tickets were sold in all?

_______ tickets

b. Noah flew 1,827 miles from Saint Louis to Los Angeles, then 7,461 miles to Sydney, Australia, and another 2,051 miles to his home in Perth. How far did he fly?

_______ miles

2. Break the number in the circle into two factors that will make it easier to multiply. Write the factors and then write the product.

a. $5 \times 16 =$ ____

is the same as

$5 \times$ ____ $\times$ ____ $=$ ____

b. $32 \times 5 =$ ____

is the same as

____ $\times$ ____ $\times 5 =$ ____

3. Write the multiplication fact you would use to figure out the division fact. Then complete the division fact.

a. $27 \div 9 =$ ____

____ $\times$ ____ $=$ ____

b. $64 \div 8 =$ ____

____ $\times$ ____ $=$ ____

c. $42 \div 7 =$ ____

____ $\times$ ____ $=$ ____

d. $54 \div 6 =$ ____

____ $\times$ ____ $=$ ____

e. $35 \div 5 =$ ____

____ $\times$ ____ $=$ ____

f. $56 \div 8 =$ ____

____ $\times$ ____ $=$ ____

# RAGING RHINO

**How do you stop a rhinoceros from charging?**

★ Figure out each of these and draw a straight line to the correct product. The line will pass through a letter. Write each letter above its matching product at the bottom of the page. Some letters appear more than once.

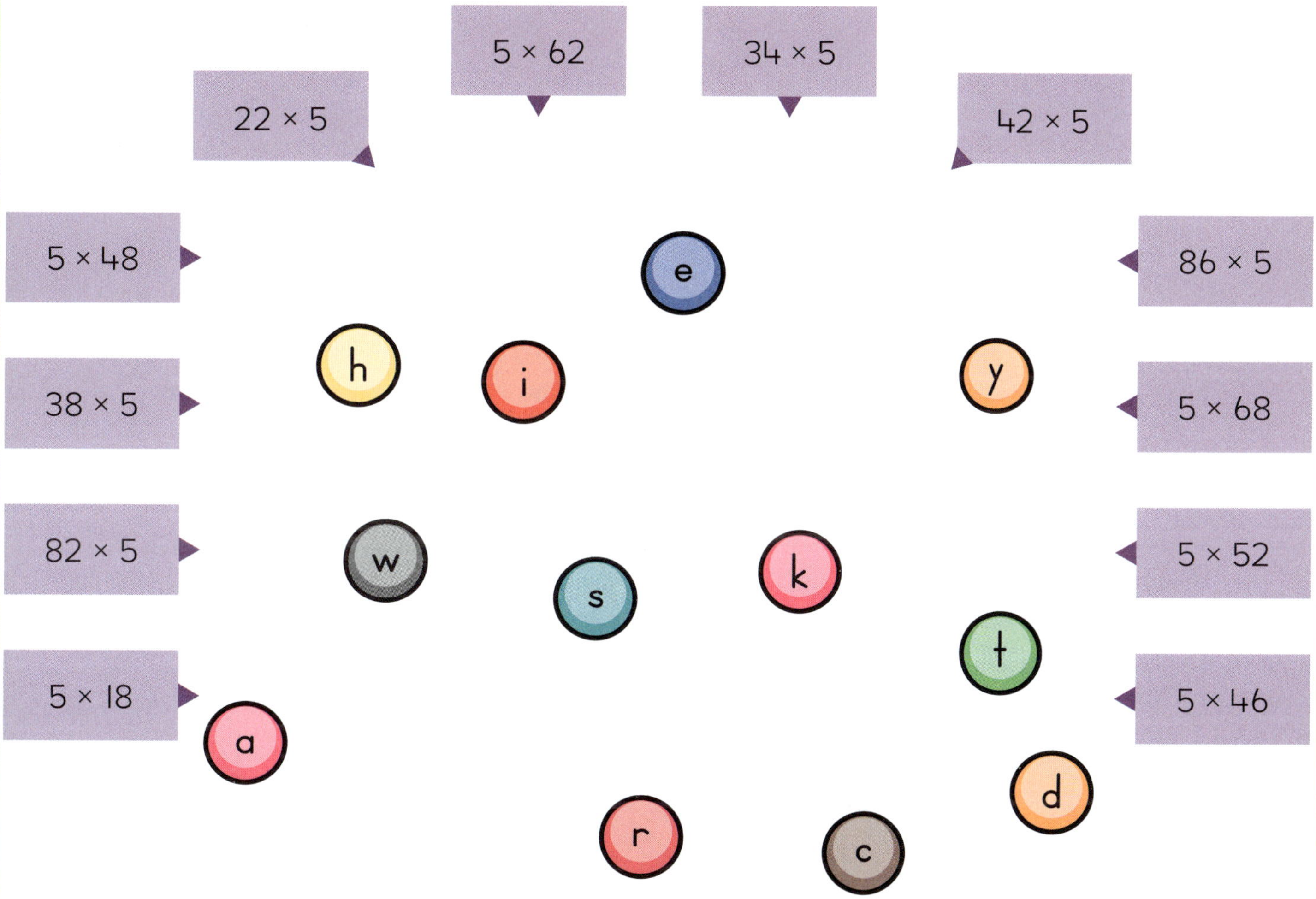


| 340 | 170 | 110 | 90 | 310 | 240 | 230 | 410 | 430 | 260 | 210 | 190 |
|---|---|---|---|---|---|---|---|---|---|---|---|
| | | | | | | | | | | | |

| ___ | ___ | ___ | ___ | | ___ | ___ | ___ | ___ | | ___ | ___ | ___ |
|---|---|---|---|---|---|---|---|---|---|---|---|---|
| 430 | 90 | 340 | 170 | | 90 | 240 | 90 | 210 | | 110 | 310 | 190 |

| ___ | ___ | ___ | ___ | ___ | ___ | | ___ | ___ | ___ | ___ |
|---|---|---|---|---|---|---|---|---|---|---|
| 230 | 410 | 170 | 260 | 310 | 430 | | 230 | 90 | 410 | 260 |

**1.** Write these times using **minutes to**.

**a.** 

_______ minutes to _____

**b.** 

_______ minutes to _____

**c.** 

_______ minutes to _____

**d.** 

_______ minutes to _____

**2.** On this number line, the distance from 0 to 1 is one whole. In each pair, loop the fraction that is **greater**. Use the number line to help.

**a.** $\frac{2}{5}$ or $\frac{3}{4}$

**b.** $\frac{6}{4}$ or $\frac{7}{5}$

**c.** $\frac{9}{4}$ or $\frac{11}{5}$

**d.** $\frac{15}{4}$ or $\frac{19}{5}$

**e.** $\frac{2}{4}$ or $\frac{3}{5}$

**f.** $\frac{9}{5}$ or $\frac{8}{4}$

**g.** $\frac{10}{4}$ or $\frac{12}{5}$

**h.** $\frac{17}{5}$ or $\frac{14}{4}$

**3.** Share these marbles into bags. Write the multiplication fact that helped you figure out the number in each bag. Then write the number left over.

**a.** 35 marbles shared into 4 bags

_____ × _____ = _______

_____ in each bag remainder _____

**b.** 26 marbles shared into 7 bags

_____ × _____ = _______

_____ in each bag remainder _____

**c.** 48 marbles shared into 5 bags

_____ × _____ = _______

_____ in each bag remainder _____

**d.** 46 marbles shared into 6 bags

_____ × _____ = _______

_____ in each bag remainder _____

**1.** Break one number into two factors that will help you figure out the product. Then write the matching multiplication sentence.

**a.** $6 \times 25 =$ ______

is the same as

____ × ____ × ____ = ________

**b.** $15 \times 8 =$ ______

is the same as

____ × ____ × ____ = ________

**c.** $45 \times 4 =$ ______

is the same as

____ × ____ × ____ = ________

**d.** $35 \times 8 =$ ______

is the same as

____ × ____ × ____ = ________

**2.** Use the standard subtraction algorithm to calculate the exact difference.

**a.**

| | Th | H | T | O |
|---|---|---|---|---|
| | 4 | 6 | 3 | 7 |
| − | 1 | 4 | 8 | 5 |
| | | | | |

**b.**

| | TTh | Th | H | T | O |
|---|---|---|---|---|---|
| | 1 | 2 | 5 | 7 | 3 |
| − | | 6 | 1 | 4 | 8 |
| | | | | | |

**c.**

| | TTh | Th | H | T | O |
|---|---|---|---|---|---|
| | 3 | 4 | 5 | 1 | 6 |
| − | | 1 | 8 | 9 | 3 |
| | | | | | |

**3.** Each large rectangle is one whole. Write how much is shaded in each rectangle. Then draw extra lines to figure out an equivalent fraction.

**a.**

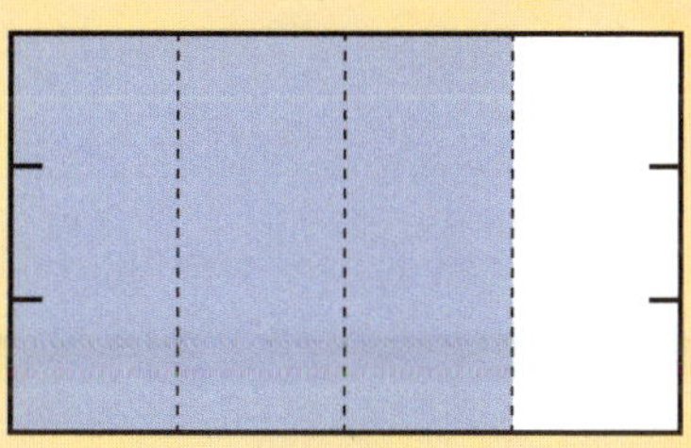

$\frac{\square}{4} = \frac{\square}{\square}$

**b.**

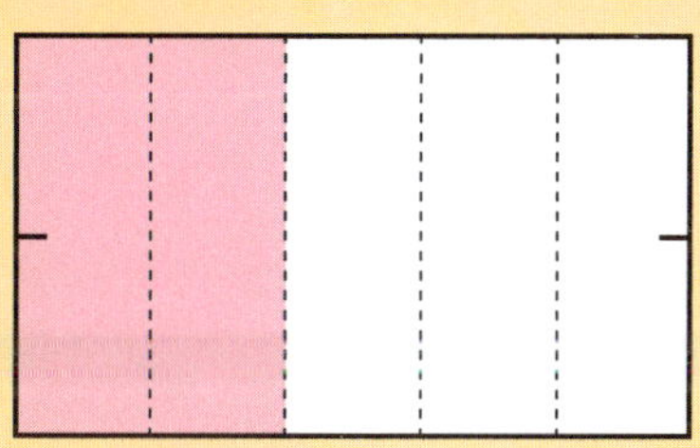

$\frac{\square}{5} = \frac{\square}{\square}$

**c.**

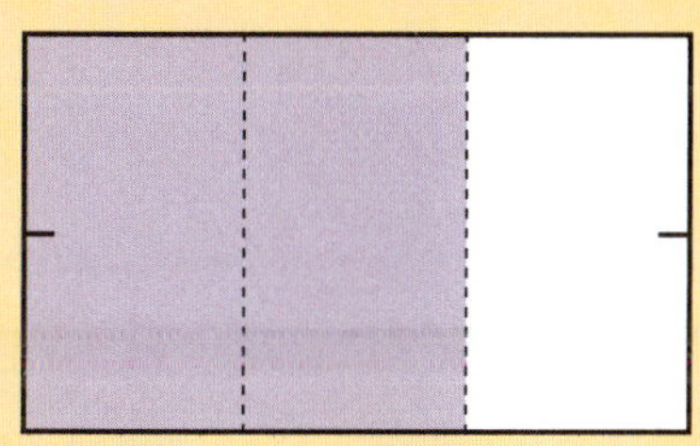

$\frac{\square}{3} = \frac{\square}{\square}$

# HOW MANY?

**How many birds can you put in an empty cage?**

★ Figure out each of these and write the answer. Then write each letter above its matching answer at the bottom of the page.

| | | |
|---|---|---|
| 190 + 470 = ____ e | 540 − 260 = ____ e | 810 − 560 = ____ f |
| 380 + 590 = ____ t | 160 + 380 = ____ s | 450 − 190 = ____ y |
| 520 − 370 = ____ h | 280 + 370 = ____ p | 460 + 280 = ____ a |
| 370 + 450 = ____ i | 560 − 380 = ____ t | 350 − 180 = ____ t |
| 250 + 670 = ____ m | 480 + 350 = ____ r | 940 − 180 = ____ o |
| 860 − 670 = ____ t | 920 − 630 = ____ o | 570 + 360 = ____ t |
| 570 + 280 = ____ t | 820 − 550 = ____ n | 930 − 350 = ____ n |
| 350 + 290 = ____ a | 450 + 360 = ____ e | |

___ ___ ___, ___ ___ ___ ___ ___ ___ ___ ___ ___
760 270 280, 640 250 970 660 830 930 150 740 850

___ ___ ' ___ ___ ___ ___ ___ ___ ___ ___ ___
820 190 ' 540 580 290 180 810 920 650 170 260

1. Double one number and halve the other. Then write the product.

a. 35 × 8 = ____

is the same as

____ × ____ = ____

b. 14 × 6 = ____

is the same as

____ × ____ = ____

c. 24 × 5 = ____

is the same as

____ × ____ = ____

d. 45 × 6 = ____

is the same as

____ × ____ = ____

e. 15 × 8 = ____

is the same as

____ × ____ = ____

f. 6 × 35 = ____

is the same as

____ × ____ = ____

2. Estimate the difference. Then use the standard subtraction algorithm to calculate the exact difference.

a. Estimate ____

| | H | T | O |
|---|---|---|---|
| | 3 | 6 | 4 |
| − | | 5 | 4 |

b. Estimate ____

| | H | T | O |
|---|---|---|---|
| | 7 | 5 | 9 |
| − | 2 | 5 | 6 |

c. Estimate ____

| | H | T | O |
|---|---|---|---|
| | 4 | 8 | 5 |
| − | 2 | 0 | 7 |

d. Estimate ____

| | H | T | O |
|---|---|---|---|
| | 2 | 0 | 7 |
| − | | 8 | 6 |

3. Complete each of these to find an equivalent fraction.

a. × ____ $\frac{2}{5} = \frac{\_\_}{10}$ × ____

b. × ____ $\frac{3}{6} = \frac{\_\_}{12}$ × ____

c. × ____ $\frac{2}{3} = \frac{\_\_}{12}$ × ____

d. × ____ $\frac{5}{8} = \frac{\_\_}{16}$ × ____

1. The distance from 0 to 1 is one whole. Write the fraction that each arrow is pointing to.

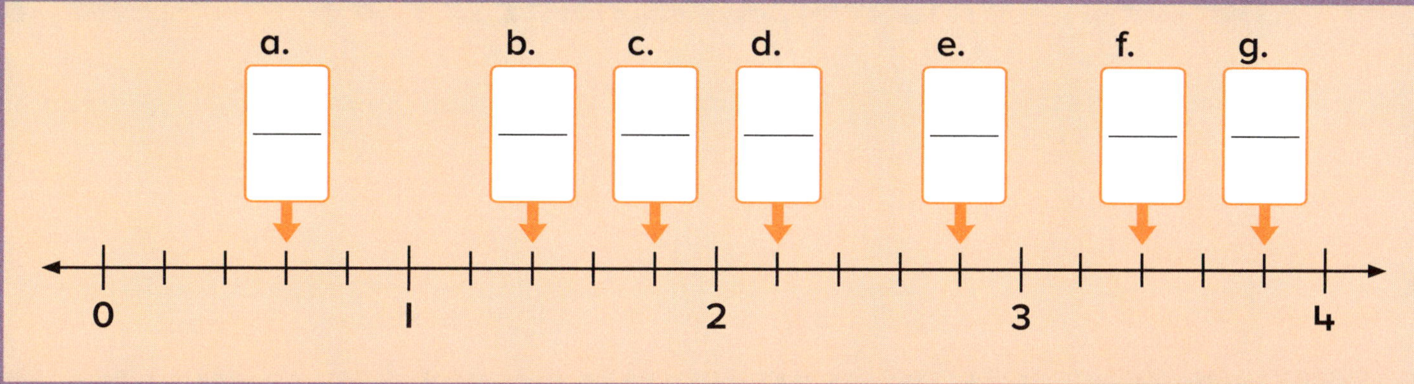


2. Solve these word problems. Show your thinking.

a. Amos wanted to buy a car valued at $17,590. He had $9,500 in his savings. The dealer gave him $8,900 for his old car. How much did he have left over?

$ ________

b. Emma wants to buy a car for $18,250. She has saved $11,125. Her parents will give her $5,000 toward the cost. How much more does she need to save?

$ ________

3. Change **one** fraction in each pair so that they have the same denominator. Then rewrite the fractions.

a.

Old $\frac{2}{3}$ $\frac{2}{6}$

New ___ ___

$\frac{\_\_}{\_\_} \times \frac{\_\_}{\_\_} = \frac{\_\_}{\_\_}$

b.

Old $\frac{5}{6}$ $\frac{14}{12}$

New ___ ___

$\frac{\_\_}{\_\_} \times \frac{\_\_}{\_\_} = \frac{\_\_}{\_\_}$

# HORKEY?

**What is the offspring of a male horse and a female donkey called?**

★ Figure out each of these and write the product. Then find each product in the panel below and color the matching letter blue.

| | | |
|---|---|---|
| $14 \times 35 =$ ____ | $18 \times 25 =$ ____ | $45 \times 12 =$ ____ |
| $12 \times 15 =$ ____ | $18 \times 55 =$ ____ | $35 \times 16 =$ ____ |
| $16 \times 15 =$ ____ | $45 \times 18 =$ ____ | $14 \times 45 =$ ____ |
| $12 \times 55 =$ ____ | $12 \times 35 =$ ____ | $15 \times 14 =$ ____ |
| $55 \times 14 =$ ____ | $45 \times 16 =$ ____ | $15 \times 18 =$ ____ |
| $16 \times 55 =$ ____ | | |

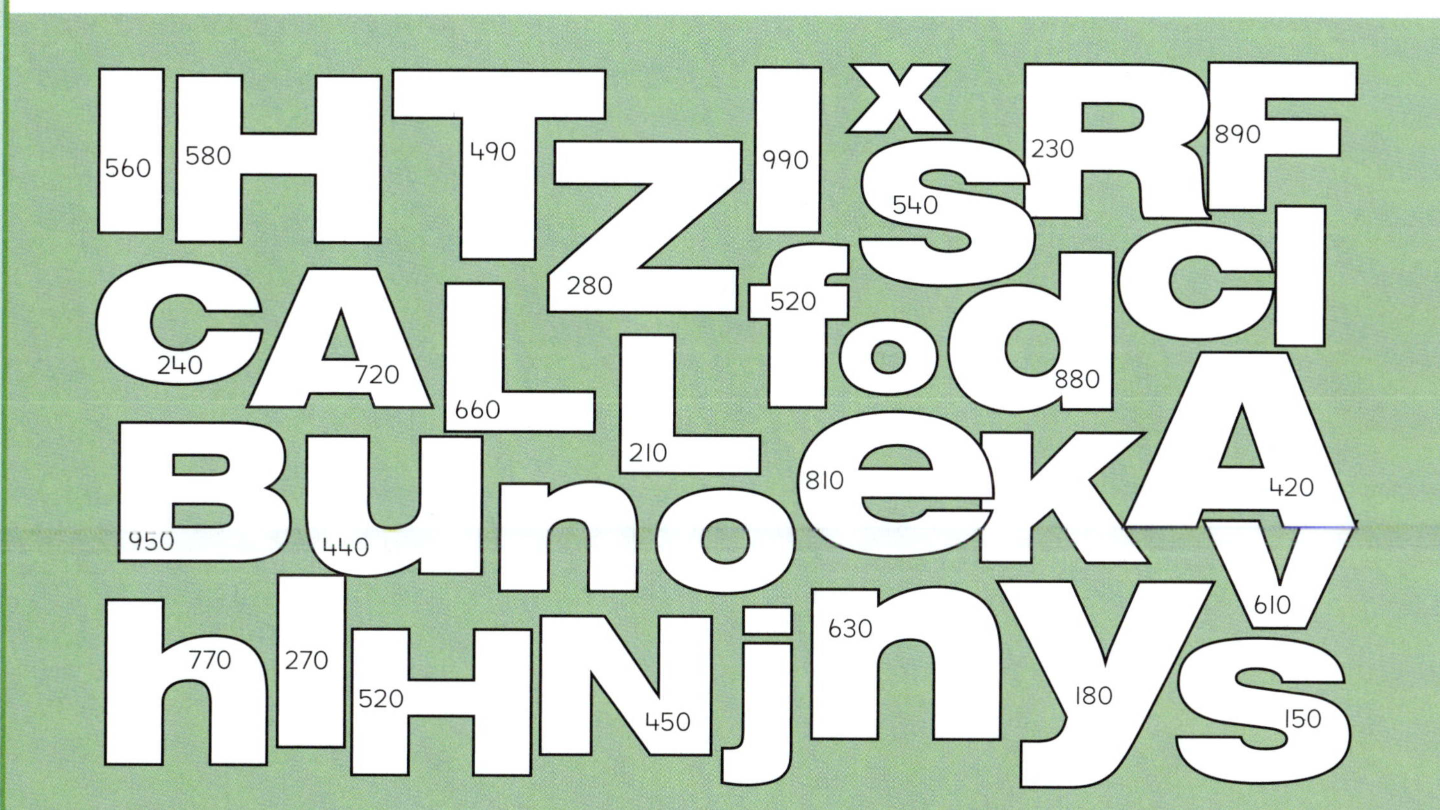

**1.** Use the standard addition algorithm to calculate the total.

**a.**
$$\begin{array}{r} 7\,5\,2\,4 \\ +\ 1\,3\,5\,0 \\ \hline \end{array}$$

**b.**
$$\begin{array}{r} 3\,0\,4\,9 \\ +\ 1\,6\,5\,3 \\ \hline \end{array}$$

**c.**
$$\begin{array}{r} 5\,1\,6\,7 \\ +\ 3\,4\,8\,5 \\ \hline \end{array}$$

**d.**
$$\begin{array}{r} 2\,4\,7\,0\,5 \\ +\ 6\,5\,3\,9 \\ \hline \end{array}$$

**2.** Use a pattern to help you complete these equations.

**a.**

42 ÷ 6 = ____

420 ÷ 6 = ____

4,200 ÷ 6 = ____

42,000 ÷ 6 = ____

**b.**

36 ÷ 9 = ____

360 ÷ 9 = ____

3,600 ÷ 9 = ____

36,000 ÷ 9 = ____

**3.** For each pair of fractions, complete the diagram to show equivalent fractions that have a common denominator. Then complete the sentence.

**a.** $\frac{2}{3}$ and $\frac{3}{4}$

A common multiple is ____

so a common denominator is ____.

$\frac{2}{3}$ = ____ (× 4 on top, × 4 on bottom)

$\frac{3}{4}$ = ____ (× 3 on top, × 3 on bottom)

**b.** $\frac{1}{3}$ and $\frac{4}{5}$

A common multiple is ____

so a common denominator is ____.

____ × ____ = ____

____ × ____ = ____

**c.** $\frac{3}{4}$ and $\frac{5}{6}$

A common multiple is ____

so a common denominator is ____.

____ × ____ = ____

____ × ____ = ____

**1.** Use the standard addition algorithm to calculate the total.

**a.**

$$\begin{array}{r} 3075 \\ 1642 \\ +\ 2115 \\ \hline \end{array}$$

**b.**

$$\begin{array}{r} 2415 \\ 26051 \\ +\ 32716 \\ \hline \end{array}$$

**c.**

$$\begin{array}{r} 42001 \\ 16950 \\ +\ 34507 \\ \hline \end{array}$$

**2.** Use a pattern to help you complete these equations.

**a.**

3 × 6 = ____

3 × 60 = ____

3 × 600 = ____

3 × 6,000 = ____

**b.**

4 × 7 = ____

4 × 70 = ____

4 × 700 = ____

4 × 7,000 = ____

**3.** Each large rectangle is one whole. Shade parts using different colors to show each fraction. Then write the total fraction that is shaded.

**a.** $\frac{2}{6} + \frac{3}{6} = \frac{\ }{\ }$

**b.** $\frac{2}{5} + \frac{1}{5} = \frac{\ }{\ }$

**c.** $\frac{3}{10} + \frac{4}{10} = \frac{\ }{\ }$

**d.** $\frac{7}{12} + \frac{3}{12} = \frac{\ }{\ }$

**e.** $\frac{4}{8} + \frac{2}{8} = \frac{\ }{\ }$

**f.** $\frac{4}{6} + \frac{1}{6} = \frac{\ }{\ }$

Time Taken:

# RACE TRACK

★ For each division fact, write the multiplication fact you would use to help you figure out the answer. Then write the answers. Complete the race as fast as you can. Use the classroom clock to time yourself.

72 ÷ 8 = ___
___ × ___ = ___

14 ÷ 2 = ___
___ × ___ = ___

40 ÷ 8 = ___
___ × ___ = ___

36 ÷ 4 = ___
___ × ___ = ___

20 ÷ 5 = ___
___ × ___ = ___

32 ÷ 4 = ___
___ × ___ = ___

16 ÷ 4 = ___
___ × ___ = ___

24 ÷ 4 = ___
___ × ___ = ___

18 ÷ 2 = ___
___ × ___ = ___

12 ÷ 6 = ___
___ × ___ = ___

48 ÷ 8 = ___
___ × ___ = ___

28 ÷ 4 = ___
___ × ___ = ___

32 ÷ 8 = ___
___ × ___ = ___

16 ÷ 2 = ___
___ × ___ = ___

14 ÷ 7 = ___
___ × ___ = ___

56 ÷ 8 = ___
___ × ___ = ___

24 ÷ 8 = ___
___ × ___ = ___

**1.** A train leaves North Station at 5 o'clock. Look at each watch below. How many minutes does each person need to wait for the train?

**a.**

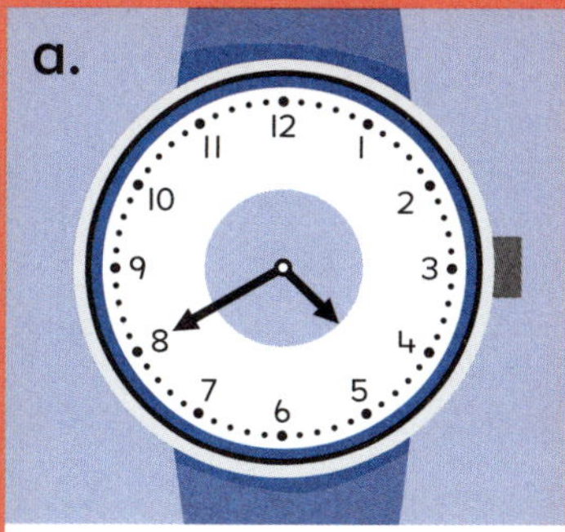

_______ minutes

**b.**

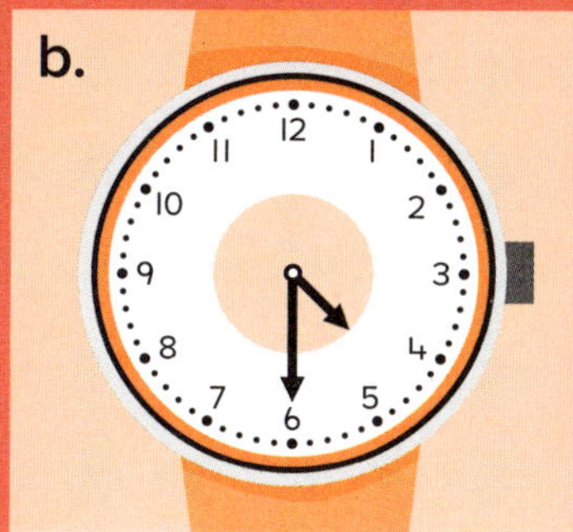

_______ minutes

**c.**

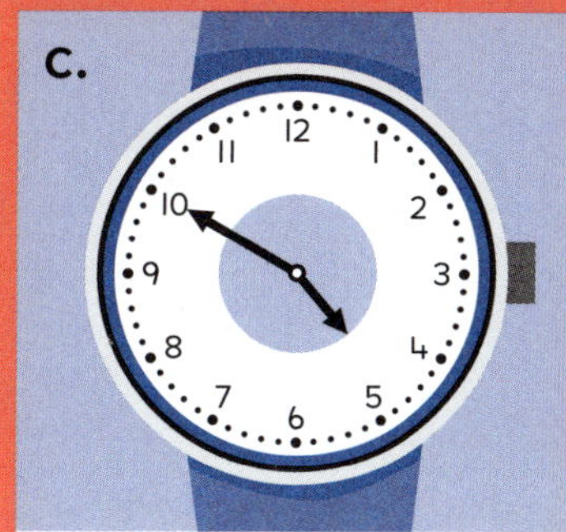

_______ minutes

**d.**

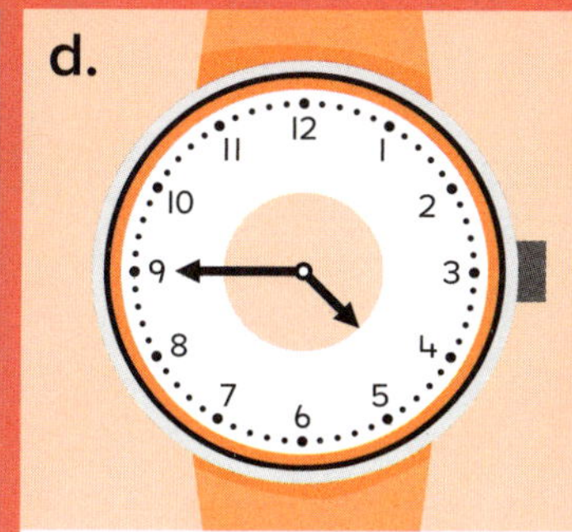

_______ minutes

**2.** Complete the thinking you could use to figure out the unknown value.

**a.**

see → $52 \div 4 =$ ?

think

4 × 10 plus 12 left over

4 × 10 plus 4 × ____

4 × ____ = ____

**b.**

see → $312 \div 6 =$ ?

think

____ × ____ plus ____ left over

____ × ____ plus ____ × ____

____ × ____ = ____

**3.** Use this number line to help you write the totals below.

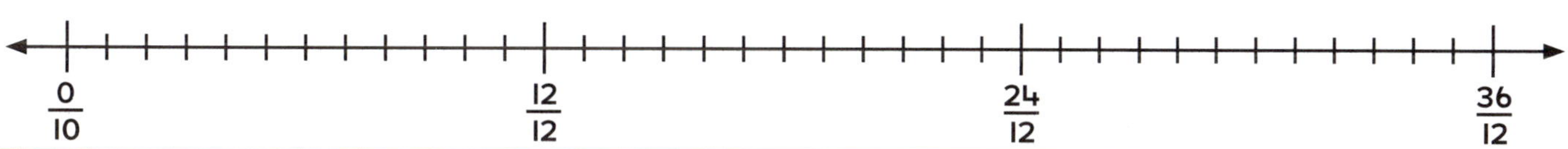


**a.** $\frac{5}{12} + \frac{3}{12} =$ ____

**b.** $\frac{9}{12} + \frac{6}{12} =$ ____

**c.** $\frac{11}{12} + \frac{5}{12} =$ ____

**d.** $\frac{14}{12} + \frac{9}{12} =$ ____

**e.** $\frac{19}{12} + \frac{12}{12} =$ ____

**f.** $\frac{21}{12} + \frac{11}{12} =$ ____

**1.** Use the standard addition algorithm to calculate the total.

**a.**

$$\begin{array}{r} 3\,6\,0\,9 \\ +\ \ \ 3\,4\,5 \\ \hline \end{array}$$

**b.**

$$\begin{array}{r} 2\,1\,6\,7 \\ +\ 3\,4\,8\,5 \\ \hline \end{array}$$

**c.**

$$\begin{array}{r} 7\,5\,6\,0 \\ +\ 1\,7\,5\,9 \\ \hline \end{array}$$

**d.**

$$\begin{array}{r} 2\,1\,0\,3\,9 \\ +\ \ \ 6\,9\,8\,0 \\ \hline \end{array}$$

**2.** Complete each of these to show an equivalent fraction.

**a.** × ____

$\frac{6}{4} = \frac{\_\_}{8}$

× ____

**b.** × ____

$\frac{2}{3} = \frac{\_\_}{6}$

× ____

**c.** × ____

$\frac{4}{6} = \frac{\_\_}{12}$

× ____

**d.** × ____

$\frac{5}{4} = \frac{\_\_}{12}$

× ____

**3.** Draw marbles in each empty bag to match the label.

**a.** Bag A has 4 times as many marbles as Bag B.

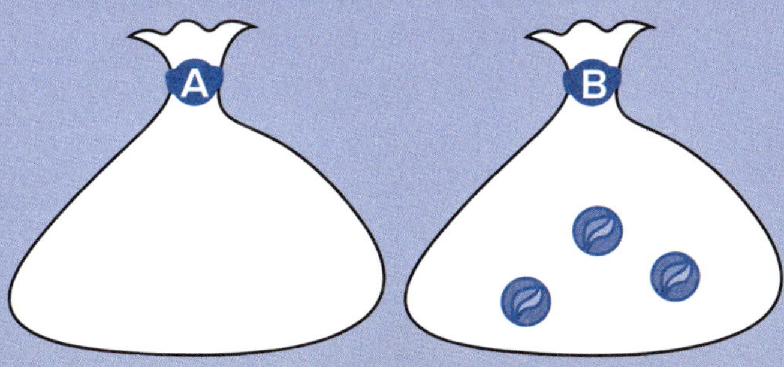


**b.** Bag C has 3 times as many marbles as Bag D.

**c.** Bag E as 2 times as many marbles as Bag F.

**d.** Bag G has 5 times as many marbles as Bag H.

# CROSS NUMBER

★ Write the total for each of these in the puzzle grid below.

**Across**

**a.** 18 + 345

**c.** 255 + 28

**e.** 537 + 19

**f.** 28 + 119

**h.** 525 + 29

**j.** 19 + 635

**l.** 423 + 29

**n.** 19 + 547

**o.** 157 + 39

**p.** 217 + 48

**Down**

**a.** 333 + 28

**b.** 328 + 29

**c.** 49 + 216

**d.** 336 + 38

**g.** 38 + 427

**i.** 527 + 48

**j.** 59 + 612

**k.** 18 + 438

**l.** 434 + 28

**m.** 19 + 256

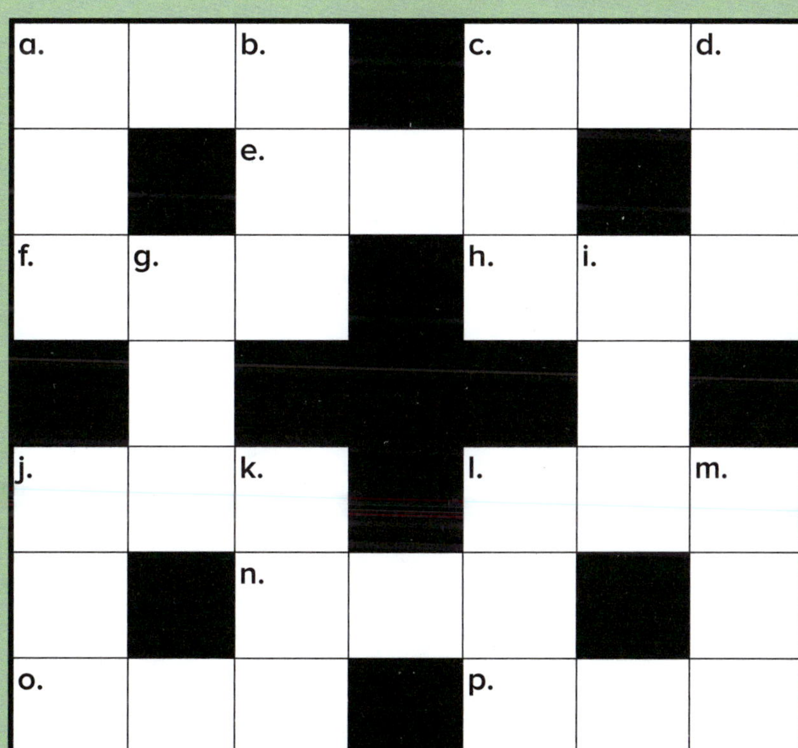

1. Break one number into two factors that will help you figure out the answer. Then write the matching multiplication sentence.

a. 6 × 25 = ______

is the same as

____ × ____ × ____ = ________

b. 15 × 8 = ______

is the same as

____ × ____ × ____ = ________

c. 45 × 4 = ______

is the same as

____ × ____ × ____ = ________

d. 35 × 8 = ______

is the same as

____ × ____ × ____ = ________

2. Write **<**, **>**, or **=** to make each sentence true. Show your thinking.

a. $\frac{2}{3}$ ◯ $\frac{3}{4}$

b. $\frac{3}{2}$ ◯ $\frac{4}{3}$

c. $\frac{5}{2}$ ◯ $\frac{10}{5}$

d. $\frac{8}{4}$ ◯ $\frac{6}{3}$

3. Write the number sentence that you would use to figure out each answer. Complete the diagram to show your thinking.

a. Ella threw 26 yd. Jack threw 8 yd farther. How far did Jack throw?

**Ella** **26**

**Jack** 8

________

b. The price in Store A is $8. The price in Store B is 3 times as much as Store A. What is the price in Store B?

**Store A** **$8**

**Store B**

________

1. Figure out the difference in value. Show your thinking.

a. $502 $314

$________

b. $727 $318

$________

c. $186 $305

$________

d. $840 $407

$________

2. Use what you know about adding fractions to complete each of these.

a. $\frac{3}{5} + \frac{1}{5} = \frac{\square}{\square}$

b. $\frac{8}{4} + \frac{7}{4} = \frac{\square}{\square}$

c. $\frac{5}{6} + \frac{7}{6} = \frac{\square}{\square}$

d. $\frac{3}{8} + \frac{11}{8} = \frac{\square}{\square}$

e. $\frac{1}{5} + \frac{6}{5} + \frac{2}{5} = \frac{\square}{\square}$

f. $\frac{7}{8} + \frac{4}{8} + \frac{11}{8} = \frac{\square}{\square}$

3. Write the number sentence that you would use to figure out each answer. Complete the diagram to show your thinking.

a. Mia is twice the height of her 2-year-old son. She is 168 cm tall. How tall is her son?

Mia | 168

Mia's Son

____________

b. Felix paid $74. Janice paid $6 less for the same item. How much did Janice pay?

Felix | $74

Janice

____________

# NAME THE SHAPE

**Two of my 3 sides are equal and 2 of my angles are equal. What shape am I?**

★ Figure out each of these and write the product. Then write each letter above its matching product at the bottom of the page.

| | | | | | |
|---|---|---|---|---|---|
| 6 × 35 = ______ | n | 4 × 65 = ______ | i | 95 × 4 = ______ | g |
| 75 × 8 = ______ | l | 8 × 55 = ______ | a | 4 × 45 = ______ | n |
| 65 × 6 = ______ | s | 35 × 8 = ______ | e | 8 × 85 = ______ | s |
| 6 × 45 = ______ | t | 4 × 35 = ______ | o | 4 × 85 = ______ | e |
| 95 × 6 = ______ | a | 75 × 6 = ______ | e | 4 × 75 = ______ | l |
| 8 × 65 = ______ | r | 45 × 8 = ______ | s | 55 × 4 = ______ | c |
| 85 × 6 = ______ | i | | | | |

| ___ | ___ | | ___ | ___ | ___ | ___ | ___ | ___ | ___ | ___ | ___ |
|---|---|---|---|---|---|---|---|---|---|---|---|
| 440 | 210 | | 510 | 360 | 140 | 390 | 220 | 340 | 600 | 450 | 680 |

| ___ | ___ | ___ | ___ | ___ | ___ | ___ | ___ |
|---|---|---|---|---|---|---|---|
| 270 | 520 | 260 | 570 | 180 | 380 | 300 | 280 |

**1.** Use the standard subtraction algorithm to calculate each difference.

**a.**

|   | 2 | 5 | 0 | 6 | 9 |
|---|---|---|---|---|---|
| − | 1 | 7 | 4 | 3 | 5 |

**b.**

|   | 3 | 7 | 1 | 2 | 5 |
|---|---|---|---|---|---|
| − | 1 | 6 | 8 | 7 | 0 |

**c.**

|   | 4 | 7 | 3 | 0 | 2 |
|---|---|---|---|---|---|
| − | 1 | 0 | 5 | 0 | 6 |

**2.** Use a protractor to measure and label the inside angles of each shape.

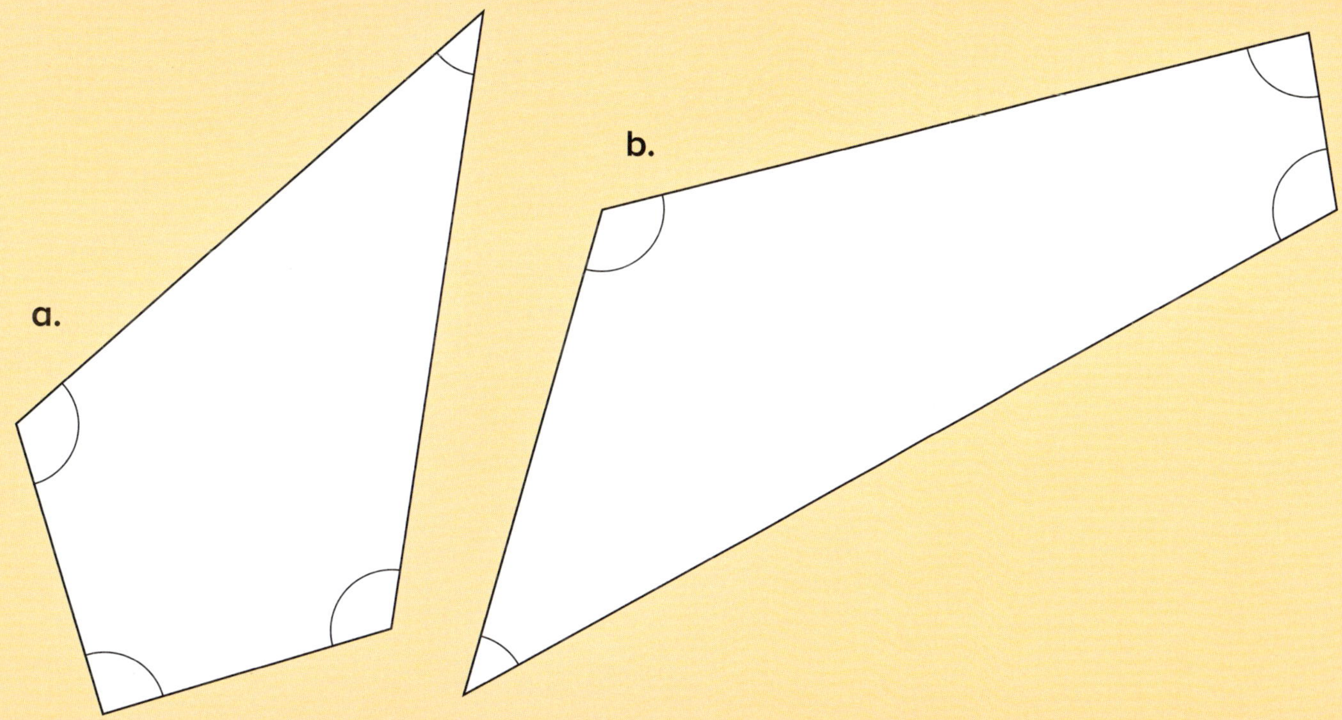


**3.** Write each whole number as a fraction. Use what you know about multiples to help you.

**a.**

$1 = \frac{\square}{2}$ $2 = \frac{\square}{2}$ $3 = \frac{\square}{2}$ $4 = \frac{\square}{2}$ $5 = \frac{\square}{2}$

**b.**

$1 = \frac{\square}{3}$ $2 = \frac{\square}{3}$ $3 = \frac{\square}{3}$ $4 = \frac{\square}{3}$ $5 = \frac{\square}{3}$

**c.**

$1 = \frac{\square}{4}$ $2 = \frac{\square}{4}$ $3 = \frac{\square}{4}$ $4 = \frac{\square}{4}$ $5 = \frac{\square}{4}$

1. Think of a multiplication fact to help you solve each division problem. Then write the answer.

a. 16 ÷ 3 = ______ r ______

b. 37 ÷ 7 = ______ r ______

c. 22 ÷ 5 = ______ r ______

d. 67 ÷ 8 = ______ r ______

e. 57 ÷ 9 = ______ r ______

f. 46 ÷ 6 = ______ r ______

2. Use a protractor to help you find these angles.

- Write **A** next to the acute angles.
- Write **R** next to the right angles.
- Write **O** next to the obtuse angles.

a.

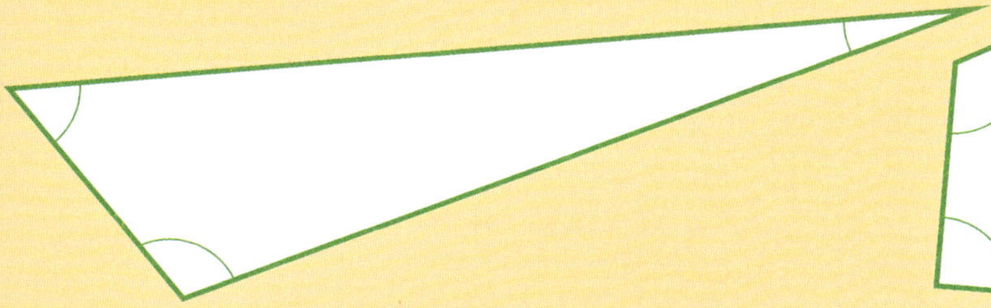

b.

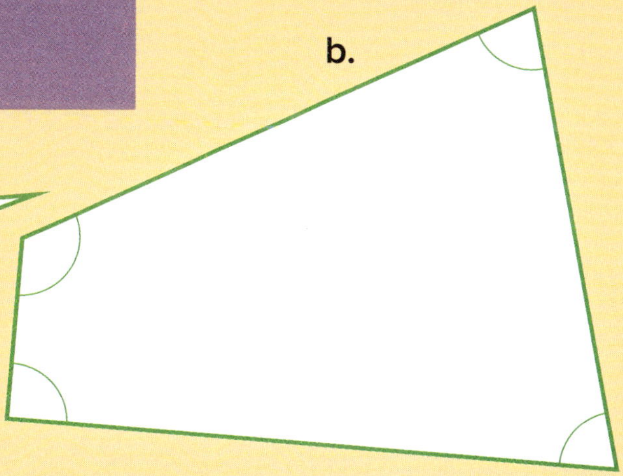

3. Each large shape is one whole. Shade the shapes to show the mixed number.

a. $3\frac{2}{3}$

b. $1\frac{4}{6}$

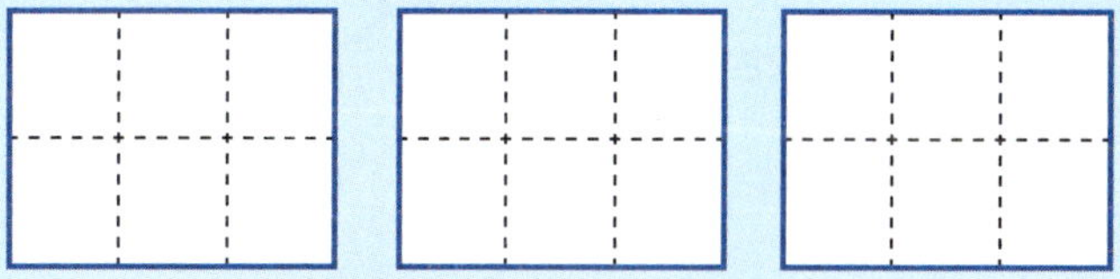

c. $2\frac{1}{4}$

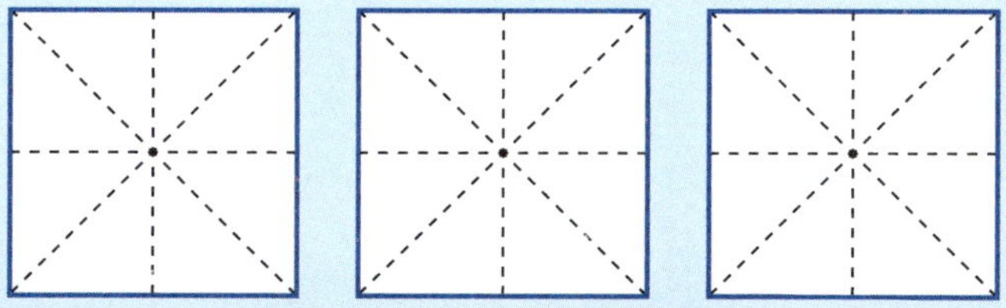

d. $3\frac{5}{8}$

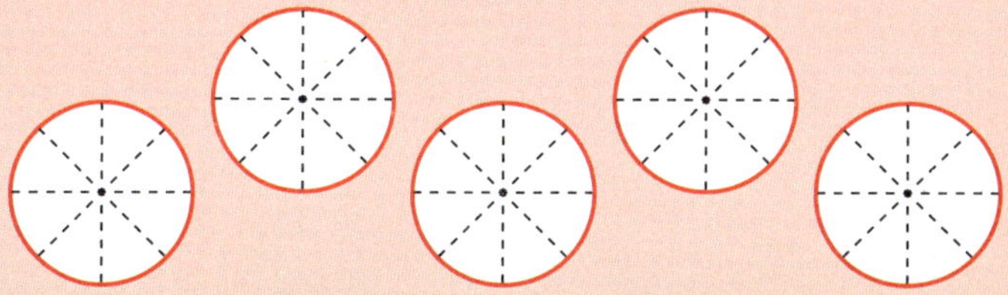

# DANCING COMPETITION

**Did you hear about the girl who does tap dancing?**

★ For each division fact, draw a straight line to a multiplication fact you could use to help you figure out the answer. The line will pass through a number and a letter. Write each letter above its matching number at the bottom of the page. Some multiplication facts are used more than once.

| Division facts |
|---|
| 45 ÷ 9 |
| 24 ÷ 6 |
| 30 ÷ 5 |
| 32 ÷ 4 |
| 28 ÷ 7 |
| 15 ÷ 3 |
| 40 ÷ 8 |
| 35 ÷ 5 |
| 30 ÷ 6 |
| 45 ÷ 5 |
| 24 ÷ 4 |
| 36 ÷ 9 |
| 28 ÷ 4 |
| 35 ÷ 7 |
| 40 ÷ 5 |
| 15 ÷ 5 |

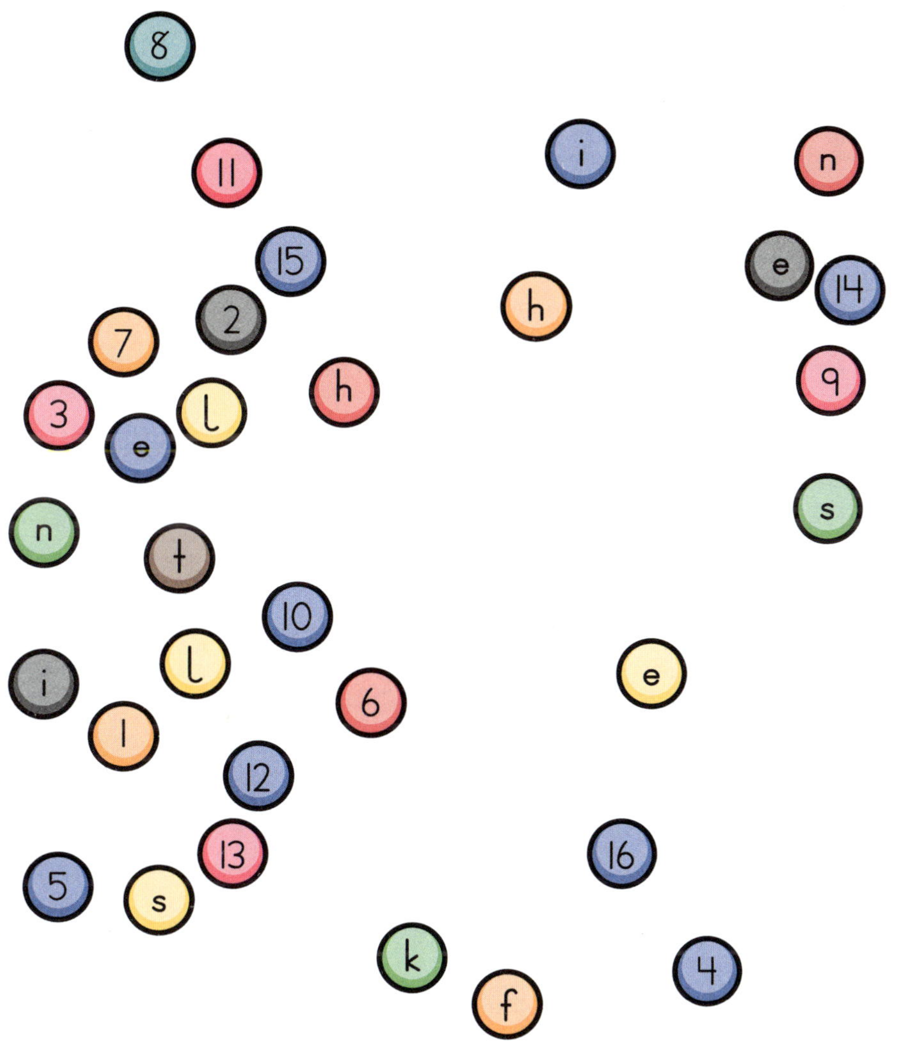


| Multiplication facts |
|---|
| 4 × 7 |
| 5 × 9 |
| 7 × 5 |
| 6 × 4 |
| 9 × 4 |
| 3 × 5 |
| 5 × 6 |
| 8 × 5 |
| 4 × 8 |

___ ___ ___   ___ ___ ___ ___   ___ ___

1 2 3   4 5 6 7   8 9

___ ___ ___   ___ ___ ___ ___

10 11 12   13 14 15 16

1. Divide the number on the expander. Then complete the equations.

a.

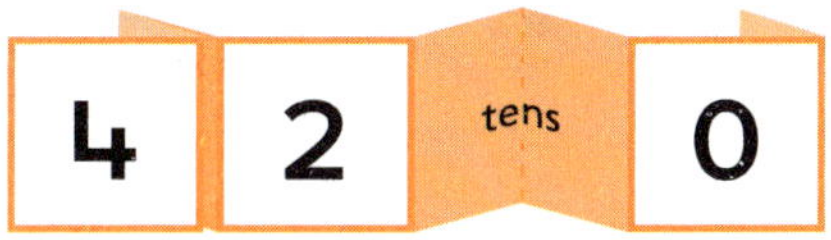


42 tens ÷ 6 = ______ tens

__________ ÷ 6 = ______

b.

72 hundreds ÷ 9 = ______ hundreds

__________ ÷ 9 = ______

2. Use the clues to calculate the size of each angle in the diagram.
Do not use a protractor. Show your thinking.

**Clues**

- Angle **AOE** is 90°.
- Angle **AOB** is 30°.
- Angle **COD** is half of Angle **COE**.
- Angle **COE** is the same size as Angle **AOB**.

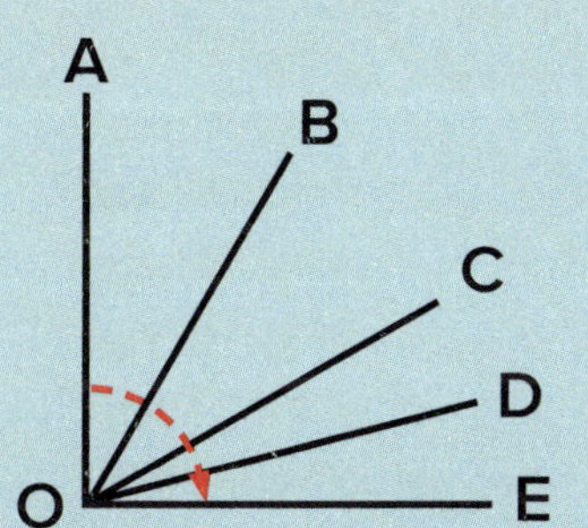


Angle **BOE** is ______°

Angle **COD** is ______°

Angle **BOC** is ______°

Angle **BOB** is ______°

3. Complete the common fractions and mixed numbers.

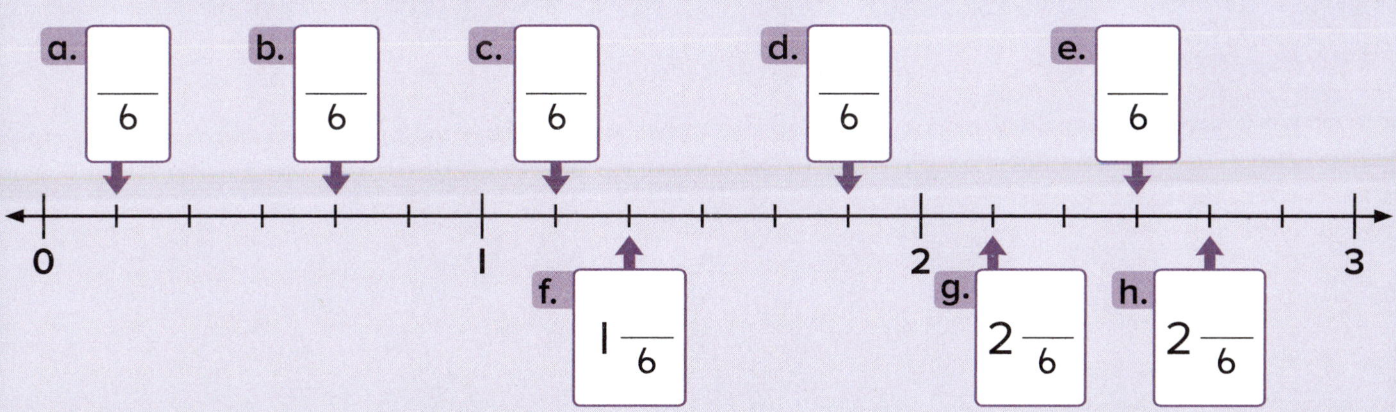

1. Use the standard addition algorithm to calculate the total.

a.
$$\begin{array}{r} 6305 \\ +\ 2874 \\ \hline \end{array}$$

b.
$$\begin{array}{r} 3079 \\ +\ 5650 \\ \hline \end{array}$$

c.
$$\begin{array}{r} 2895 \\ +\ 6144 \\ \hline \end{array}$$

d.
$$\begin{array}{r} 23480 \\ +\ 12795 \\ \hline \end{array}$$

2. Write the number sentence that you would use to figure out each answer. Complete the diagram to show your thinking.

a. Eva has 15 baseball cards. Lilly has 4 times as many as Eva. How many baseball cards does Lilly have?

Eva | 15 |

Lilly

b. James has $15. Damon has $4 more than James. How much money does Damon have?

James | 15 |

Damon

3. Color the tens part red and the ones part blue. Then write each product. Add the two partial products and write the total.

a. 3 × 36

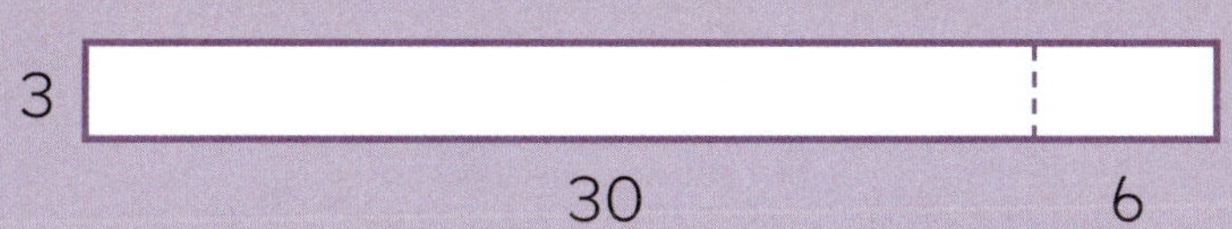


3 × ______ = ______

3 × ______ = ______

Total ______

b. 6 × 47

6 × ______ = ______

6 × ______ = ______

Total ______

# WHERE AM I?

**What does an elephant have that no other animals have?**

★ Figure out each of these and write the answer. Find the answer in the grid below and cross out the letter above. Then write the remaining letters at the bottom of the page.

860 − 310 = ______

420 + 220 = ______

680 − 470 = ______

520 + 260 = ______

780 − 610 = ______

340 + 230 = ______

750 − 410 = ______

320 + 160 = ______

570 − 250 = ______

120 + 470 = ______

670 − 450 = ______

310 + 460 = ______

550 − 240 = ______

280 + 610 = ______

770 − 520 = ______

430 + 220 = ______

660 − 420 = ______

440 + 350 = ______

980 − 360 = ______

✻

| B | L | O | N | G | E | A | R |
|---|---|---|---|---|---|---|---|
| 150 | 780 | 890 | 550 | 790 | 320 | 760 | 250 |
| S | B | G | R | E | Y | B | E |
| 220 | 580 | 240 | 770 | 480 | 470 | 210 | 560 |
| L | E | P | T | R | U | N | K |
| 230 | 180 | 330 | 640 | 650 | 310 | 620 | 340 |
| H | A | N | D | T | U | S | K |
| 490 | 630 | 260 | 170 | 660 | 590 | 460 | 570 |

| | | | |
|---|---|---|---|

| | | | | | | | | |
|---|---|---|---|---|---|---|---|---|

1. Use the standard subtraction algorithm to calculate the difference.

a.
```
  6 4 5
- 3 8 2
```

b.
```
  7 6 1
- 2 4 9
```

c.
```
  5 1 3
- 1 8 4
```

d.
```
  8 0 7
- 3 1 9
```

2. Write the number sentence that you would use to figure out each answer. Complete the diagram to show your thinking.

a. Koda bought a ticket to Ocean World theme park for $45. Zoe used a coupon and saved $15. How much did she pay?

| **Koda** | **$45** |
|---|---|
| **Zoe** | |

______________

b. Jacob is 24 years old. He is 6 times the age of his daughter Kimie. How old is Kimie?

| **Jacob** | **24** |
|---|---|
| **Kimie** | |

______________

3. Complete each number sentence. Then write the total of the three products.

a. **3 × 254**

3 × 200 = 600

3 × 50 = ______

3 × 4 = ______

Total ______

b. **6 × 183**

6 × 100 = ______

6 × 80 = ______

6 × 3 = ______

Total ______

c. **7 × 438**

7 × 400 = ______

7 × 30 = ______

7 × 8 = ______

Total ______

**1.** Use a pattern to help complete these equations.

**a.**

$3 \times 4 =$ ____

$30 \times 4 =$ ____

$300 \times 4 =$ ____

$3{,}000 \times 4 =$ ____

**b.**

$9 \times 6 =$ ____

$90 \times 6 =$ ____

$900 \times 6 =$ ____

$9{,}000 \times 6 =$ ____

**2.** Write each fraction as a whole number.

**a.** $\frac{5}{5} =$ ____ $\frac{10}{5} =$ ____ $\frac{15}{5} =$ ____ $\frac{20}{5} =$ ____ $\frac{25}{5} =$ ____

**b.** $\frac{12}{6} =$ ____ $\frac{18}{6} =$ ____ $\frac{24}{6} =$ ____ $\frac{36}{6} =$ ____ $\frac{54}{6} =$ ____

**3.** Label the dimensions for each part and write number sentences to figure out each partial product. Then add the partial products to figure out the total.

**a.**

**4 × 3,265**

____ × ____ = ____

____ × ____ = ____

____ × ____ = ____

____ × ____ = ____

Total ____

**b.**

**3 × 2,584**

____ × ____ = ____

____ × ____ = ____

____ × ____ = ____

____ × ____ = ____

Total ____

Time Taken:

# RACE TRACK

★ For each division fact, write the multiplication fact you would use to help figure out the answer. Then write the answers. Complete the race as fast as you can. Use the classroom clock to time yourself.

35 ÷ 5 = ___
___ × ___ = ___

36 ÷ 9 = ___
___ × ___ = ___

15 ÷ 5 = ___
___ × ___ = ___

40 ÷ 8 = ___
___ × ___ = ___

32 ÷ 4 = ___
___ × ___ = ___

45 ÷ 5 = ___
___ × ___ = ___

24 ÷ 6 = ___
___ × ___ = ___

36 ÷ 4 = ___
___ × ___ = ___

30 ÷ 6 = ___
___ × ___ = ___

16 ÷ 4 = ___
___ × ___ = ___

28 ÷ 4 = ___
___ × ___ = ___

35 ÷ 7 = ___
___ × ___ = ___

30 ÷ 5 = ___
___ × ___ = ___

28 ÷ 7 = ___
___ × ___ = ___

15 ÷ 3 = ___
___ × ___ = ___

24 ÷ 4 = ___
___ × ___ = ___

20 ÷ 5 = ___
___ × ___ = ___

**1.** Use a pattern to help complete these equations.

**a.**

35 ÷ 5 = ______

350 ÷ 5 = ______

3,500 ÷ 5 = ______

35,000 ÷ 5 = ______

**b.**

28 ÷ 7 = ______

280 ÷ 7 = ______

2,800 ÷ 7 = ______

28,000 ÷ 7 = ______

**2.** Add these mixed numbers. Show your thinking on the number line.

**a.** $2\frac{1}{4} + 2\frac{2}{4} =$ ______

**b.** $2\frac{3}{8} + 1\frac{4}{8} =$ ______

**3.** Write the expressions to match.
Then use the working space to figure out each product.

**a.** ______ × ______ = ______

**b.** ______ × ______ = ______

Working Space

1. Read each problem carefully. Figure out the answer. You can draw a picture to help.

**a.** Liam bought $\frac{3}{4}$ lb of potatoes and $\frac{2}{4}$ lb of onions. What was the total mass?

$\frac{\square}{\square}$ lb

**b.** A recipe needs $\frac{3}{8}$ qt of juice and $\frac{6}{8}$ qt of water. How much liquid is used?

$\frac{\square}{\square}$ qt

2. Count on from one mixed number to figure out the total. Draw jumps on the number line to show your thinking.

**a.** $1\frac{3}{5} + 1\frac{3}{5} =$ ________

**b.** $1\frac{3}{4} + 2\frac{3}{4} =$ ________

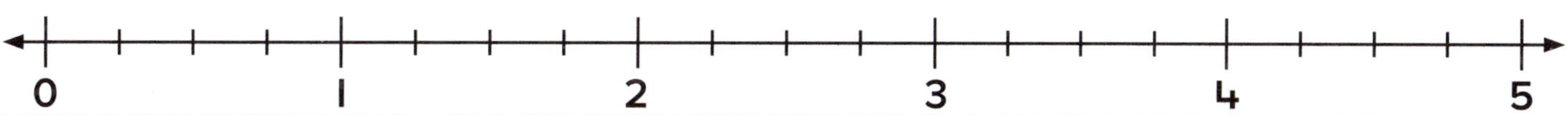


3. Write a multiplication sentence to show each part. Then write the total of the four partial products.

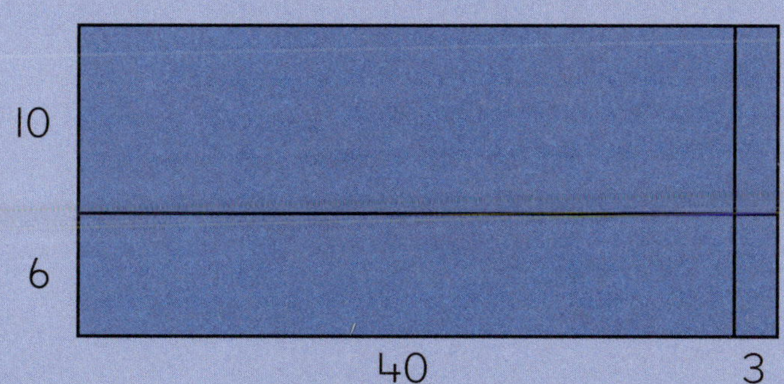


**16 × 43**

________ × ________ = ________

________ × ________ = ________

________ × ________ = ________

________ × ________ = ________

Total ________

# BELIEVE IT OR NOT!

★ To reveal a fun fact, figure out each of these and write the product. Then write each letter above its matching product at the bottom of the page. Some letters appear more than once.

| | | | | | | | | |
|---|---|---|---|---|---|---|---|---|
| $6 \times 32 =$ | ______ | r | $8 \times 45 =$ | ______ | o | $9 \times 48 =$ | ______ | l |
| $17 \times 5 =$ | ______ | u | $26 \times 4 =$ | ______ | i | $33 \times 9 =$ | ______ | a |
| $5 \times 23 =$ | ______ | q | $37 \times 7 =$ | ______ | p | $44 \times 9 =$ | ______ | f |
| $32 \times 5 =$ | ______ | t | $4 \times 29 =$ | ______ | n | $7 \times 42 =$ | ______ | m |
| $9 \times 28 =$ | ______ | e | $37 \times 8 =$ | ______ | d | $27 \times 4 =$ | ______ | c |
| $34 \times 6 =$ | ______ | s | $53 \times 6 =$ | ______ | h | | | |

| | | | | | | | | | |
|---|---|---|---|---|---|---|---|---|---|
| ______ | ______ | ______ | ______ | ______ | ______ | ______ | ______ | ______ | ______ |
| 294 | 360 | 204 | 115 | 85 | 104 | 160 | 360 | 252 | 204 |

| | | | | | |
|---|---|---|---|---|---|
| ______ | ______ | ______ | ______ | ______ | ______ |
| 259 | 192 | 252 | 396 | 252 | 192 |

| | | | | | | | |
|---|---|---|---|---|---|---|---|
| ______ | ______ | ______ | ______ | ______ | ______ | ______ | ______ |
| 108 | 318 | 104 | 432 | 296 | 192 | 252 | 116 |

| | | | | | | | |
|---|---|---|---|---|---|---|---|
| ______ | ______ | ______ | ______ | ______ | ______ | ______ | ______ |
| 160 | 360 | 297 | 296 | 85 | 432 | 160 | 204 |

**1.** In each shape, color parts to show the first fraction. Then draw more lines and complete the diagram to show an equivalent fraction.

**a.**

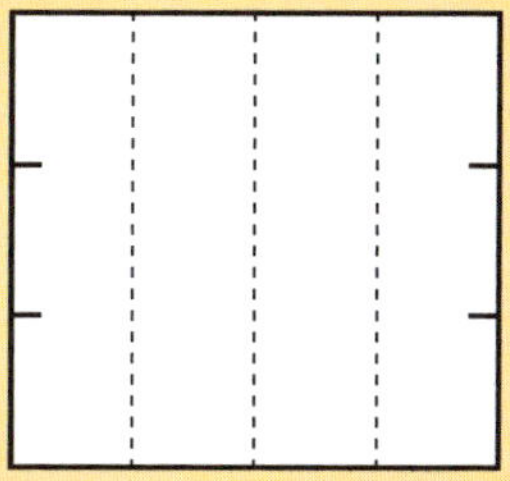

$\frac{3}{4}$ is the same as $\frac{\_\_\_}{12}$ (× ___, × ___)

**b.**

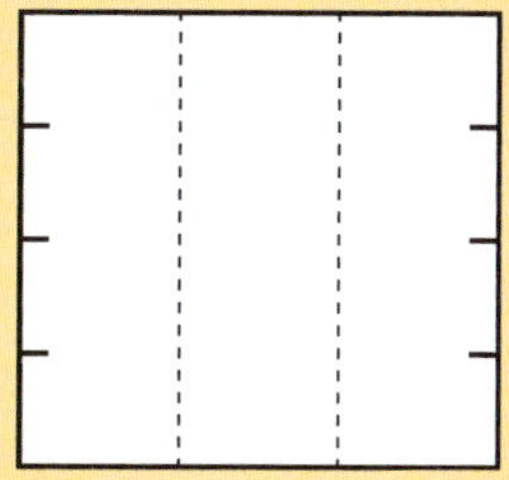

$\frac{2}{3}$ is the same as $\frac{\_\_\_}{12}$ (× ___, × ___)

**2.** Figure out the answer to each problem. Show your thinking.

**a.** Zola rides her bike $2\frac{1}{2}$ miles to school each day. How far is the trip to and from school?

__________ miles

**b.** Two hoses are joined together. One is $12\frac{7}{12}$ feet long and the other is $10\frac{6}{12}$ feet. What is the total length?

__________ feet

**3.** Figure out each product. Show your thinking.

**a.** 18 × 26 = __________

**b.** 33 × 29 = __________

**c.** 37 × 41 = __________

1. Write the whole number that is closest to each fraction. Show your thinking.

a. _____ is closest to $\frac{18}{5}$

b. _____ is closest to $\frac{16}{3}$

c. _____ is closest to $\frac{29}{8}$

2. Figure out each product. Show your thinking.

a. $3 \times 5{,}615 =$ _____

b. $5 \times 3{,}192 =$ _____

3. Draw extra beads on the abacus to match the number on the expander.

a.

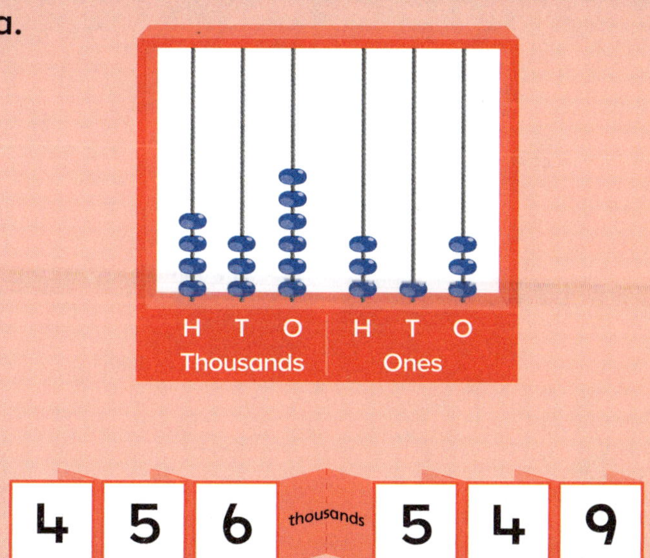


b.

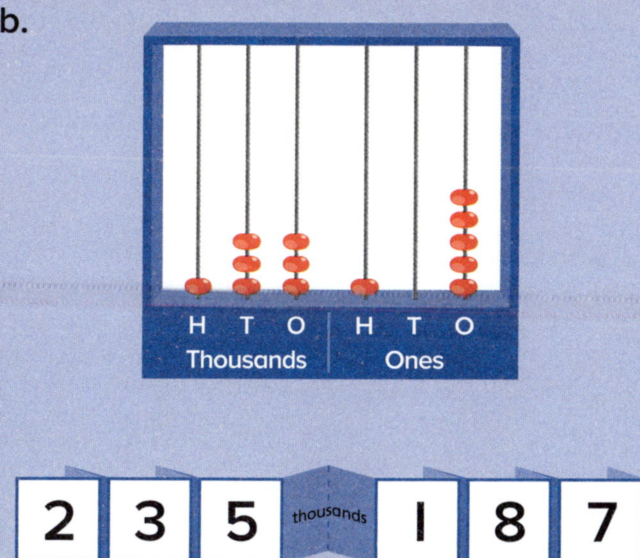

# TOO HOT!

**Why is "B" the hottest letter in the alphabet?**

★ Figure out each of these and write the answer. Then write each letter above its matching answer at the bottom of the page.

| | | | | | | | |
|---|---|---|---|---|---|---|---|
| 180 + 350 = ______ | a | 620 − 380 = ______ | e | 560 − 280 = ______ | i |
| 250 + 490 = ______ | i | 470 + 460 = ______ | l | 610 − 450 = ______ | b |
| 640 − 380 = ______ | b | 270 + 390 = ______ | e | 420 + 290 = ______ | o |
| 820 − 530 = ______ | a | 360 − 180 = ______ | u | 560 + 250 = ______ | t |
| 490 + 380 = ______ | m | 940 − 370 = ______ | l | 730 − 560 = ______ | i |
| 570 + 340 = ______ | s | 380 + 560 = ______ | c | 830 − 470 = ______ | e |
| 940 − 180 = ______ | s | 170 + 260 = ______ | o | 370 + 270 = ______ | k |

______ ______ ______ ______ ______ ______ ______    ______ ______
260 360 940 530 180 760 240    170 810

______ ______ ______ ______ ______    ______ ______ ______    ______ ______ ______ ______
870 290 640 660 910    430 280 570    160 710 740 930

1. Add these mixed numbers. Show your thinking.

a. $4\frac{4}{6} + 2\frac{1}{6} =$ ________

b. $3\frac{2}{5} + 1\frac{1}{5} =$ ________

c. $3\frac{2}{8} + 4\frac{5}{8} =$ ________

2. Write a multiplication sentence to show each part.
Then write the total of the four partial products.

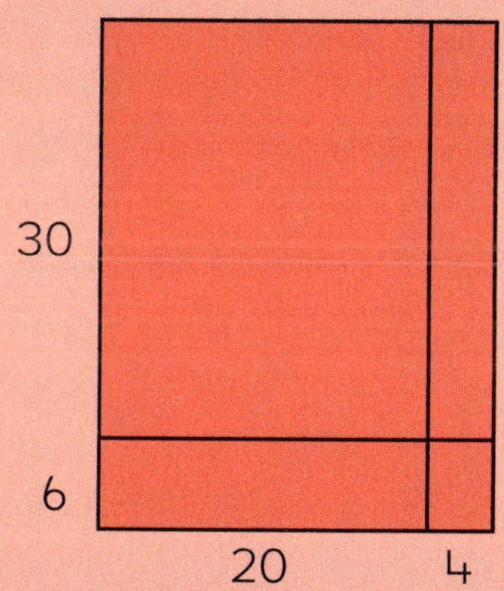


**36 × 24**

______ × ______ = ________

______ × ______ = ________

______ × ______ = ________

______ × ______ = ________

Total ________

3. Write the matching number on the expander.
Then write the number in words.

a.

4 hundred thousands

1 ten thousand

________________________________

________________________________

b.

9 hundred thousands

6 ten thousands

________________________________

________________________________

1. Split each mixed number into whole numbers and fractions before adding. Then write the total. Show your thinking.

a. $4\frac{1}{4} + 3\frac{2}{4} =$ ______

b. $3\frac{5}{6} + 2\frac{2}{6} =$ ______

c. $5\frac{6}{8} + 3\frac{5}{8} =$ ______

2. Figure out each product. Show your thinking.

a. $23 \times 47 =$ ______

b. $48 \times 33 =$ ______

c. $34 \times 56 =$ ______

3. Calculate the values and write the matching number on the expander. Then write the number in words.

a.

6 × 100,000

9 × 100

4 × 1,000

______

______

b.

3 × 1,000

9 × 10

4 × 100,000

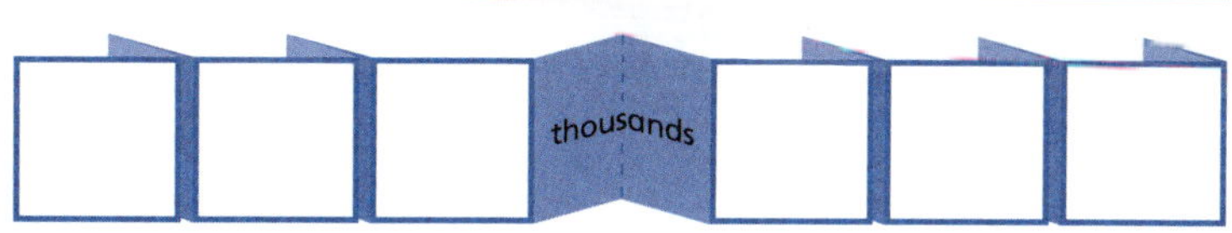


______

______

# SPIDER LEGS

**Which spiders have the shortest legs?**

★ For each division fact, draw a straight line to the correct answer. The line will pass through a number and a letter. Write the letter above its matching number at the bottom of the page. Some answers are used more than once.

**Answers**

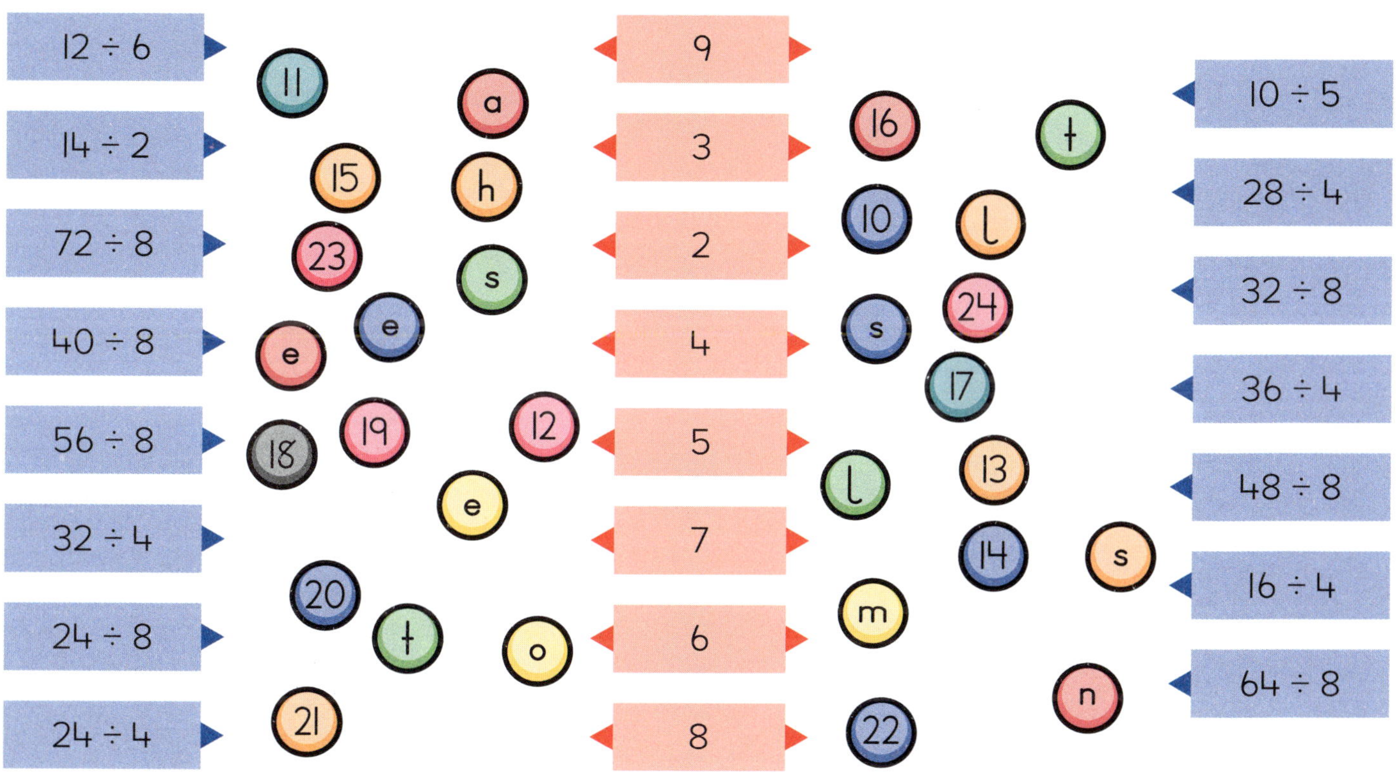


10 11 12 13 14 15 16 17 18 19 20

21 22 23 24

**1.** Use the standard addition algorithm to calculate the total.

**a.**
$$\begin{array}{r} 32121 \\ +\ 24985 \\ \hline \end{array}$$

**b.**
$$\begin{array}{r} 45799 \\ +\ 12075 \\ \hline \end{array}$$

**c.**
$$\begin{array}{r} 3501 \\ 2673 \\ +\ 2684 \\ \hline \end{array}$$

**d.**
$$\begin{array}{r} 1599 \\ 3045 \\ +\ 2166 \\ \hline \end{array}$$

**2.** Use this number line to help you write the differences.

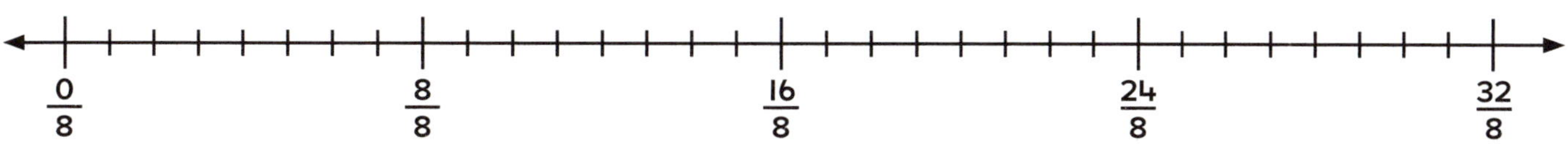


**a.** $\frac{20}{8} - \frac{5}{8} = \frac{\square}{\square}$

**b.** $\frac{\square}{\square} = \frac{17}{8} - \frac{11}{8}$

**c.** $\frac{27}{8} - \frac{10}{8} = \frac{\square}{\square}$

**d.** $\frac{\square}{\square} = \frac{12}{8} - \frac{7}{8}$

**e.** $\frac{31}{8} - \frac{20}{8} = \frac{\square}{\square}$

**f.** $\frac{\square}{\square} = \frac{18}{8} - 2$

**3. a.** Draw a line to connect each number to its location on the number line.

294,000 299,000 301,000 306,000

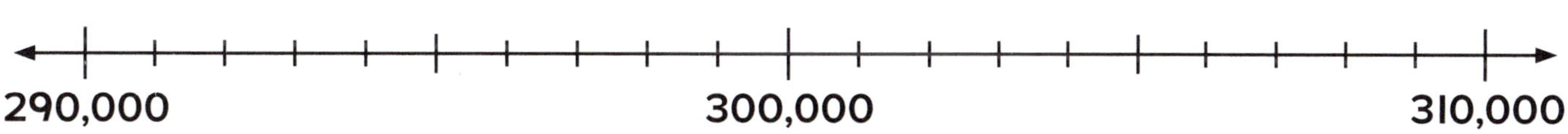


291,000 296,000 305,000 309,000

**b.** Write the number that is shown by each arrow.

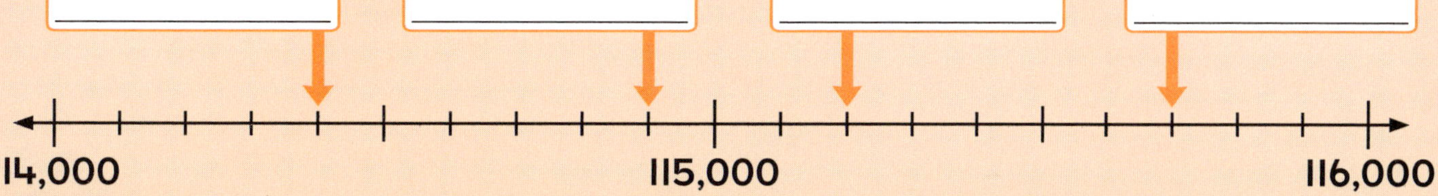

1. Look at the diagram. Use the clues to calculate the size of each angle. Do not use a protractor. Show your thinking.

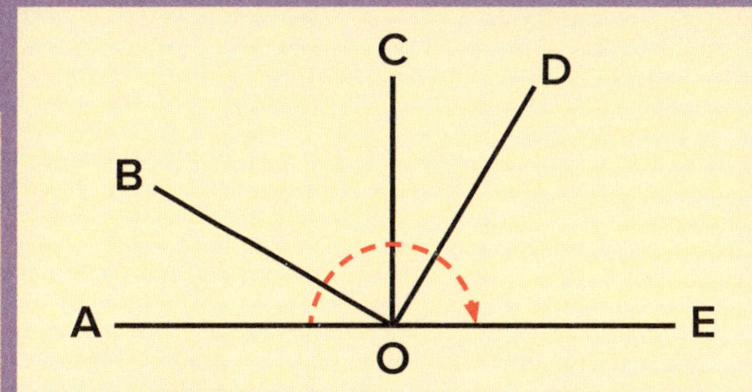


**Clues**

- Angle **AOC** is 90°.
- Angle **AOD** is 120°.
- Angle **AOB** is 30°.
- Angle **AOE** is 180°.

| Angle **BOC** is ________ ° | Angle **COD** is ________ ° | Angle **BOE** is ________ ° |
|---|---|---|

2. Calculate the difference. Draw jumps on the number line to show your thinking.

a. $3\frac{4}{6} - 1\frac{2}{6} =$ ☐

b. $3\frac{4}{5} - 1\frac{1}{5} =$ ☐

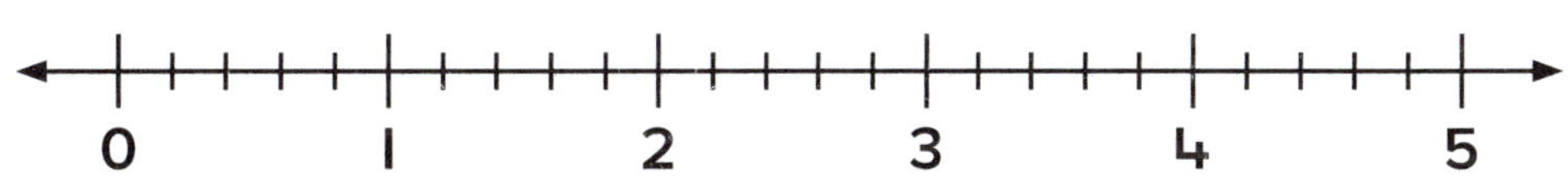


3. Write the new numbers that would be shown on each of these odometers.

a.

| | 406715 | 218452 | 047599 |
|---|---|---|---|
| 100 more miles | ________ | ________ | ________ |

b.

| | 406715 | 218452 | 047599 |
|---|---|---|---|
| 10,000 more miles | ________ | ________ | ________ |

# WOW!

★ To reveal a fun fact, figure out each of these. Then write each letter above its matching product at the bottom of the page.

| | |
|---|---|
| 2 × 6 × 3 = ______ | e |
| 7 × 8 × 2 = ______ | e |
| 9 × 2 × 8 = ______ | a |
| 7 × 5 × 3 = ______ | o |
| 8 × 4 × 5 = ______ | i |
| 6 × 5 × 7 = ______ | s |
| 2 × 7 × 9 = ______ | t |
| 5 × 6 × 4 = ______ | v |
| 4 × 2 × 6 = ______ | u |
| 2 × 8 × 6 = ______ | t |
| 3 × 6 × 5 = ______ | e |
| 3 × 5 × 9 = ______ | h |
| 8 × 5 × 6 = ______ | o |
| 9 × 6 × 2 = ______ | q |
| 9 × 2 × 9 = ______ | h |
| 4 × 7 × 2 = ______ | s |
| 8 × 7 × 5 = ______ | t |
| 5 × 2 × 6 = ______ | e |
| 6 × 2 × 7 = ______ | m |

| ___ | ___ | ___ | ___ | ___ | ___ | ___ | ___ | ___ | ___ |
|---|---|---|---|---|---|---|---|---|---|
| 84 | 105 | 56 | 108 | 48 | 160 | 280 | 240 | 36 | 210 |

| ___ | ___ | ___ | ___ | | ___ | ___ | ___ | ___ | ___ |
|---|---|---|---|---|---|---|---|---|---|
| 135 | 144 | 120 | 90 | | 96 | 112 | 60 | 126 | 162 |

**1.** Use the standard subtraction algorithm to calculate the difference.

**a.**

```
   3 9 6 4 5
 − 1 6 3 8 7
```

**b.**

```
   7 5 2 8 1
 − 2 7 1 5 9
```

**c.**

```
   6 3 0 5 2
 − 3 8 0 6 9
```

**d.**

```
   4 4 6 1 2
 − 3 9 5 3 5
```

**2.** Calculate the difference. Draw jumps on the number line to show your thinking.

**a.** $4\frac{1}{4} - 2\frac{3}{4} = \square$

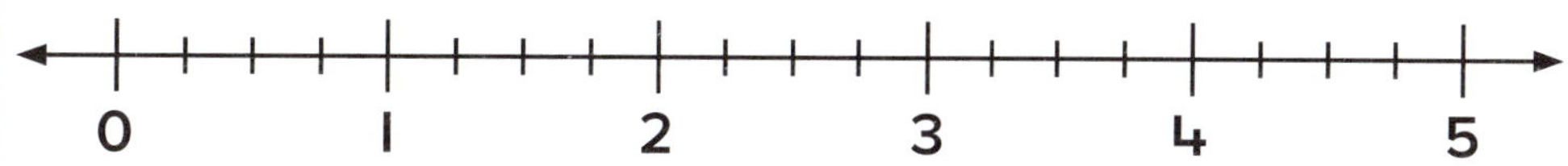


**b.** $3\frac{2}{5} - 1\frac{4}{5} = \square$

**3.** Write the missing lengths in meters and centimeters. Then draw lines to show where the other lengths are located on the measuring tape.

1 m 83 cm

____ m ______ cm

2 m 26 cm

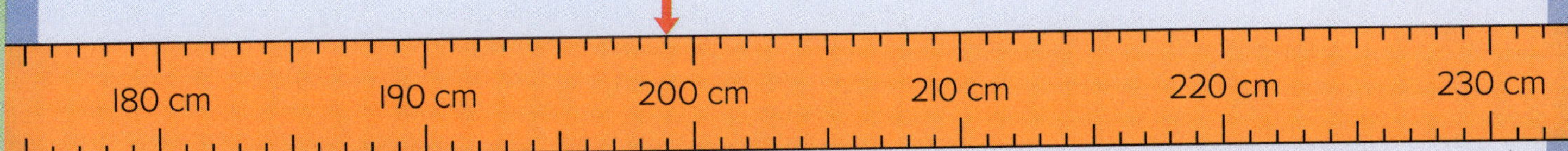


1 m 79 cm

____ m ______ cm

4 m 87 cm

____ m ______ cm

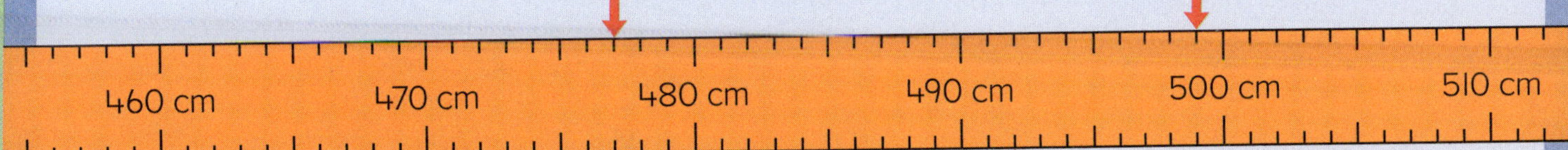


____ m ______ cm

4 m 81 cm

____ m ______ cm

5 m 9 cm

**1.** Use the standard algorithm to calculate these.

**a.**
```
   3 6 1 5
   4 0 7 2
 + 1 4 3 1
 ---------
```

**b.**
```
   2 8 0 0
   6 1 7 5
 + 3 4 2 9
 ---------
```

**c.**
```
   2 7 4 8 2
 - 1 3 6 4 5
 -----------
```

**d.**
```
   3 5 7 2 9
 -   8 3 7 5
 -----------
```

**2.** Write the missing lengths in meters and centimeters. Then draw lines to show where the other lengths are located on the measuring tape.

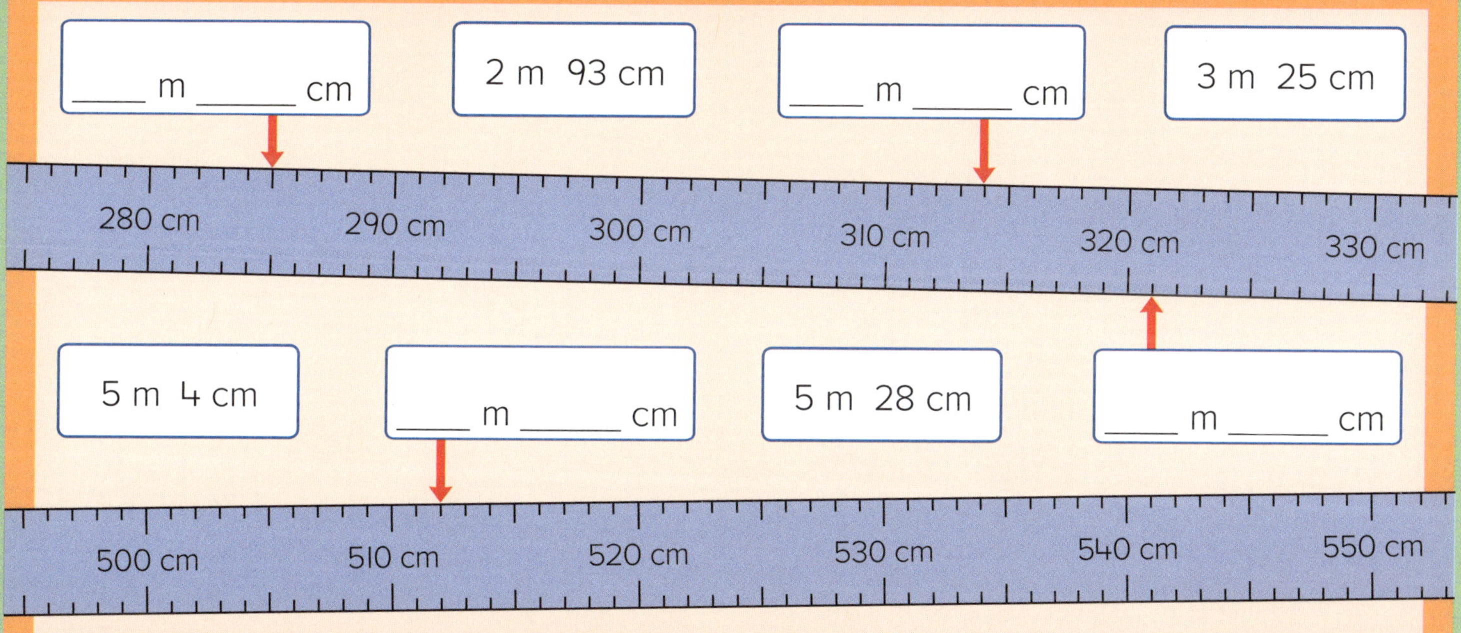


**3.** Calculate the area of each rectangle. Show your thinking.

**a.**

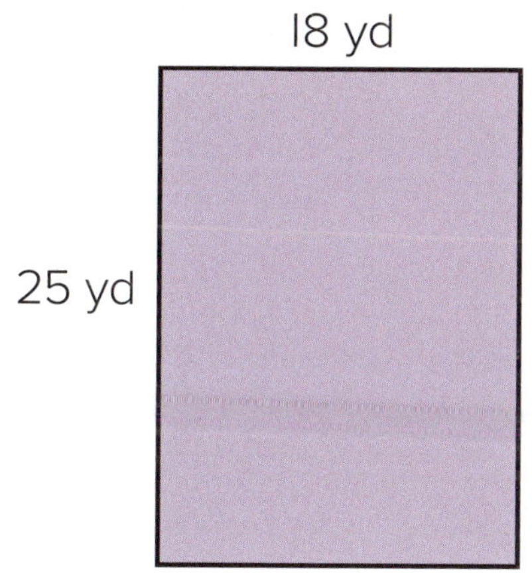


Area ________ $yd^2$

**b.**

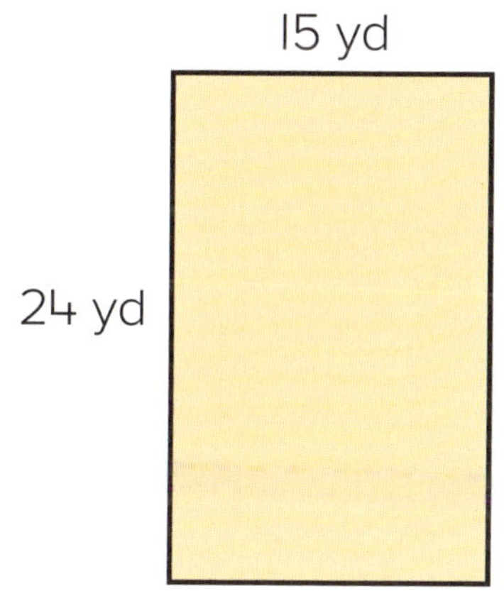


Area ________ $yd^2$

# CROSSBREEDING

**What do you get if you cross a porcupine with a giraffe?**

★ Figure out each of these and write the total. Then find each total in the puzzle below and color its matching letter red.

326 + 218 = ______

336 + 258 = ______

463 + 108 = ______

326 + 148 = ______

235 + 346 = ______

218 + 356 = ______

338 + 257 = ______

436 + 227 = ______

369 + 423 = ______

417 + 105 = ______

358 + 214 = ______

337 + 146 = ______

248 + 537 = ______

148 + 726 = ______

265 + 326 = ______

116 + 237 = ______

528 + 259 = ______

648 + 255 = ______

359 + 237 = ______

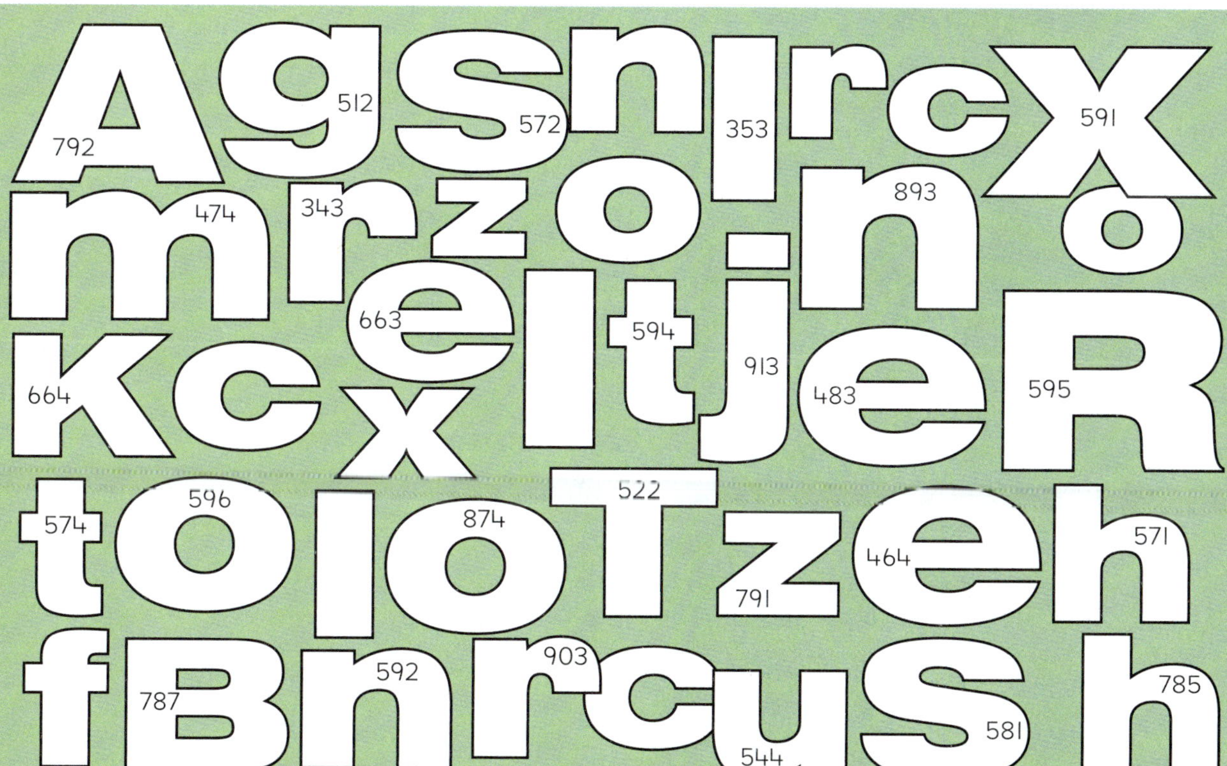

1. Write the dimensions around the rectangle. Write each partial product. Then add the partial products to figure out the total.

a. 3 × 378

____ × ____ = ____

____ × ____ = ____

____ × ____ = ____

Total ____

b. 4 × 2,186

____ × ____ = ____

____ × ____ = ____

____ × ____ = ____

____ × ____ = ____

Total ____

2. a. Measure and write the dimensions of this sticker in millimeters. Then complete the statements below.

____ mm

____ mm

b. ____ millimeters is the same as 3 centimeters.

c. ____ millimeters is the same as 30 centimeters.

d. ____ millimeters is the same as 100 centimeters.

3. Calculate the area of the shaded part.

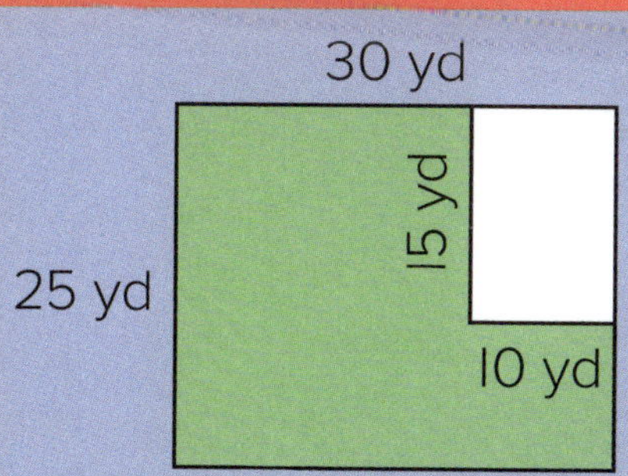


Area ____ $yd^2$

Working Space

1. Figure out each product. Show your thinking.

a. $32 \times 74 =$ ______

b. $28 \times 36 =$ ______

c. $39 \times 53 =$ ______

2. Complete each of these.

a. 1 meter **is the same length as** ______ millimeters

b. 5 meters **is the same length as** ______ millimeters

c. 10 meters **is the same length as** ______ millimeters

d. 15 meters **is the same length as** ______ millimeters

e. $\frac{1}{2}$ meter **is the same length as** ______ millimeters

f. $12\frac{1}{2}$ meters **is the same length as** ______ millimeters

3. Complete the calculation to figure out the perimeter of each rectangle.

a.

$P = 2 \times (33 + 17)$

$P = 2 \times$ ______

$P =$ ______ yd

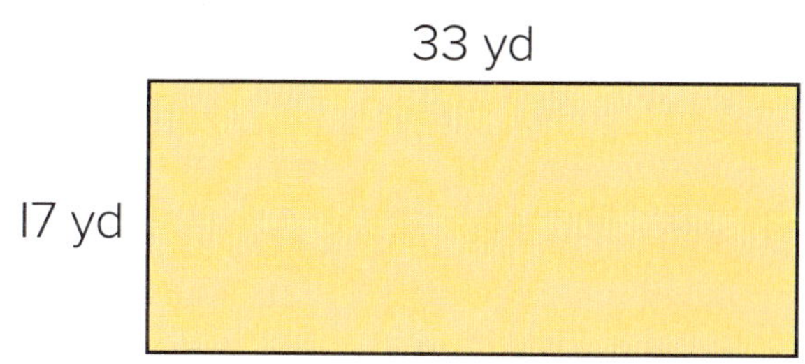


b.

$P = 2 \times (28 + 16)$

$P = 2 \times$ ______

$P =$ ______ yd

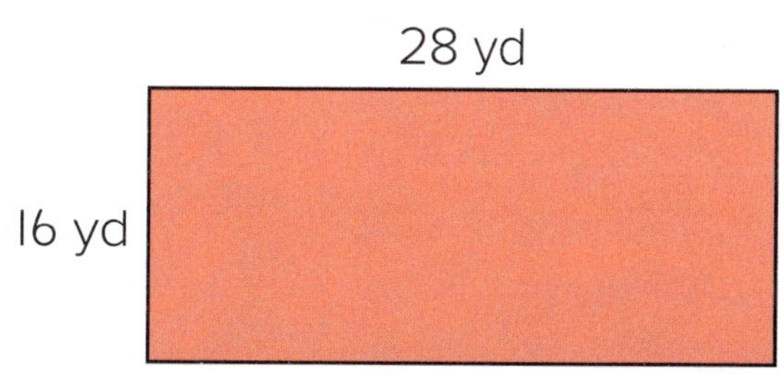

Time Taken:

# RACE TRACK

★ Figure out and write the answers to these as fast as you can. Use the classroom clock to time yourself.

$7 \div 1 =$ ____ $12 \div 4 =$ ____ $56 \div 8 =$ ____

$16 \div 8 =$ ____ $27 \div 3 =$ ____ $63 \div 9 =$ ____

$30 \div 6 =$ ____ $49 \div 7 =$ ____ $24 \div 8 =$ ____

$45 \div 5 =$ ____ $18 \div 2 =$ ____ $8 \div 1 =$ ____

$64 \div 8 =$ ____ $81 \div 9 =$ ____ $36 \div 4 =$ ____

$42 \div 7 =$ ____ $12 \div 6 =$ ____ $15 \div 3 =$ ____

$72 \div 8 =$ ____ $54 \div 9 =$ ____ finish

**1.** Solve these word problems. Show your thinking.

**a.** A stadium grandstand has 46 rows with 52 seats in each row. How many seats are there?

_______ seats

**b.** A sports club bought 26 jerseys at \$47 each. What was the total cost?

\$ _______

**2.** Complete the missing numbers in this table to show equivalent lengths.

| Millimeters (mm) | Centimeters (cm) | Meters (m) |
|---|---|---|
| 1,000 | | |
| | | $\frac{1}{2}$ |
| | 150 | |
| | | 5 |

**3.** Each large shape is one whole.
Shade each shape to match the equation then write the product.

**a.**

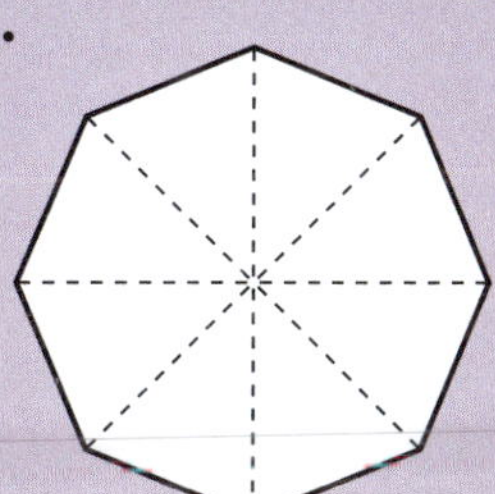

$3 \times \frac{2}{8} =$ ☐

**b.**

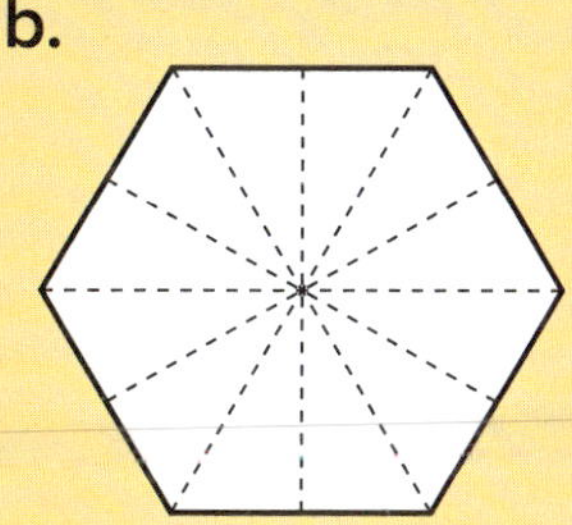

$2 \times \frac{3}{12} =$ ☐

**c.**

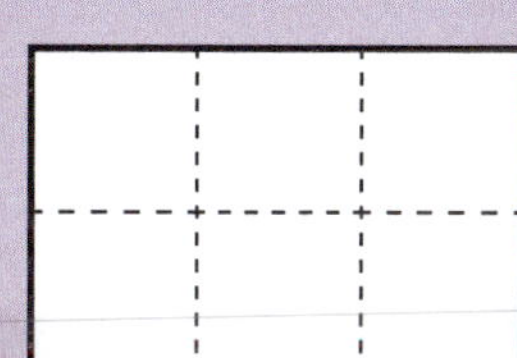

$2 \times \frac{2}{6} =$ ☐

**d.**

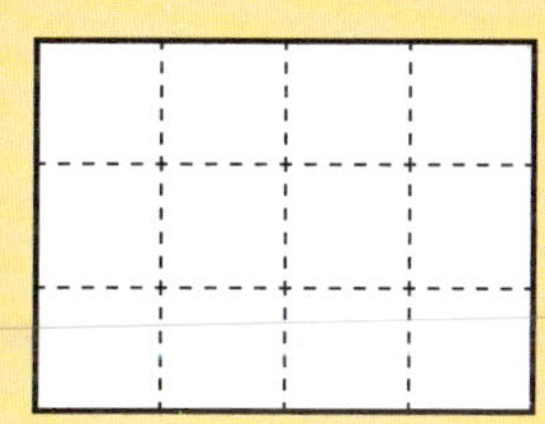

$4 \times \frac{2}{12} =$ ☐

1. Calculate the difference. Show your thinking.

a. $4\frac{4}{5} - 1\frac{1}{5} = \square$

b. $5\frac{8}{12} - 3\frac{2}{12} = \square$

c. $3\frac{6}{8} - 1\frac{2}{8} = \square$

2. Complete the missing numbers in this table to show equivalent lengths.

| Centimeters (cm) | Meters (m) | Kilometers (km) |
|---|---|---|
| | 1,000 | |
| | 100 | |
| | | $\frac{1}{2}$ |
| | 10 | |

3. The distance between each whole number is one whole.
Draw jumps to show each equation. Then write the product.

a. $3 \times \frac{4}{6} = \square$

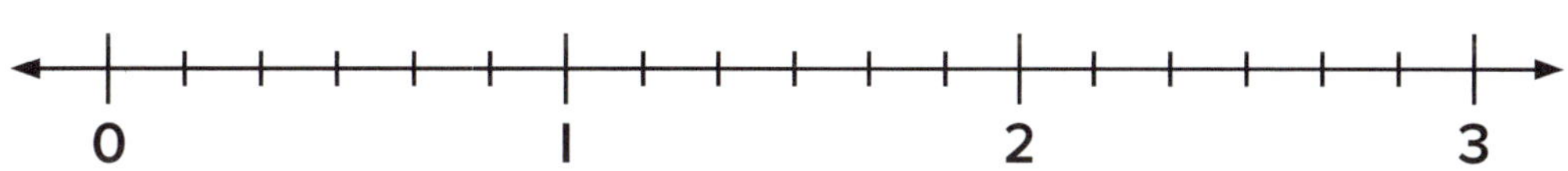


b. $6 \times \frac{2}{3} = \square$

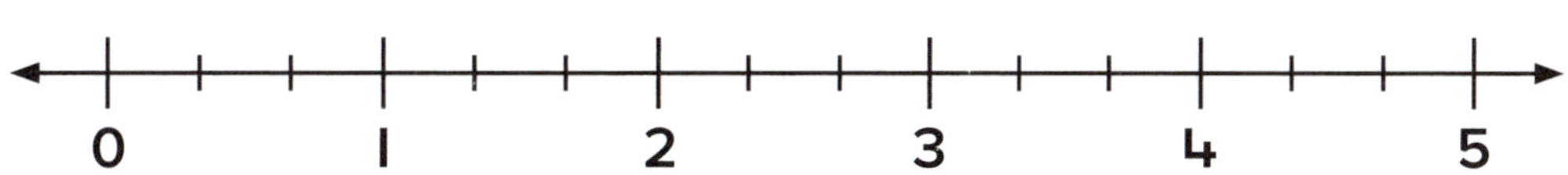

# MONSTER MEDICINE

**What do you give a seasick monster?**

★ Figure out each of these and write the product. Find each product in the grid below and cross out the letter above. Then write the remaining letters at the bottom of the page.

| | | |
|---|---|---|
| 3 × 46 = ______ | 53 × 4 = ______ | 6 × 48 = ______ |
| 54 × 7 = ______ | 6 × 54 = ______ | 49 × 6 = ______ |
| 47 × 5 = ______ | 63 × 7 = ______ | 5 × 63 = ______ |
| 39 × 8 = ______ | 5 × 36 = ______ | 6 × 64 = ______ |
| 7 × 47 = ______ | 4 × 48 = ______ | 74 × 5 = ______ |
| 4 × 36 = ______ | 3 × 76 = ______ | 54 × 3 = ______ |
| 6 × 28 = ______ | 65 × 3 = ______ | |

✻

| L | O | L | L | I | E | S | O | D | A |
|---|---|---|---|---|---|---|---|---|---|
| 138 | 324 | 202 | 212 | 168 | 180 | 288 | 421 | 162 | 378 |
| T | A | B | L | E | T | S | O | F | T |
| 344 | 144 | 235 | 195 | 294 | 228 | 170 | 334 | 302 | 315 |
| D | R | I | N | K | G | L | O | O | M |
| 384 | 364 | 192 | 441 | 312 | 329 | 370 | 148 | 172 | 185 |

☐☐☐☐ ☐☐ ☐☐☐☐

**1.** Calculate the difference. Show your thinking.

**a.** $4\frac{3}{8} - 1\frac{5}{8} = \square$

**b.** $\square = 5\frac{1}{4} - 3\frac{3}{4}$

**c.** $7\frac{2}{6} - 4\frac{3}{6} = \square$

**2.** Solve each problem. Show your thinking and be sure to use the correct units in your answer.

**a.** A plumber needs 3 meters of pipe. He joins one length that is 156 cm and another that is 2,050 mm. What is the combined length?

_______ cm

**b.** Luis rode $1\frac{3}{4}$ km to the store, another 400 m to the park then $1\frac{1}{4}$ km home. How far did he ride?

_______ m

**3.** Complete the missing numbers to calculate each product.

**a.** $4 \times 2\frac{1}{3}$

$(4 \times \square) + (4 \times \frac{\square}{\square})$

$\square + \frac{\square}{\square} = \square$

**b.** $5 \times 3\frac{2}{6}$

$(5 \times \square) + (5 \times \frac{\square}{\square})$

$\square + \frac{\square}{\square} = \square$

**1.** Each large shape is one whole. Complete each equation to show how much is shaded. Write the total as a mixed number.

**a.**

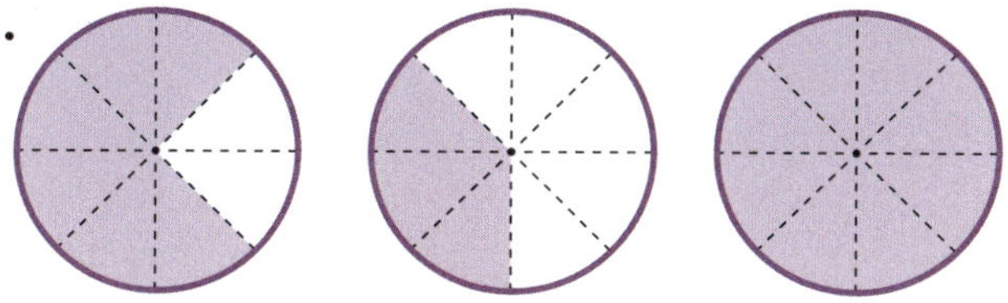

___ + ___ + ___ = ______

**b.**

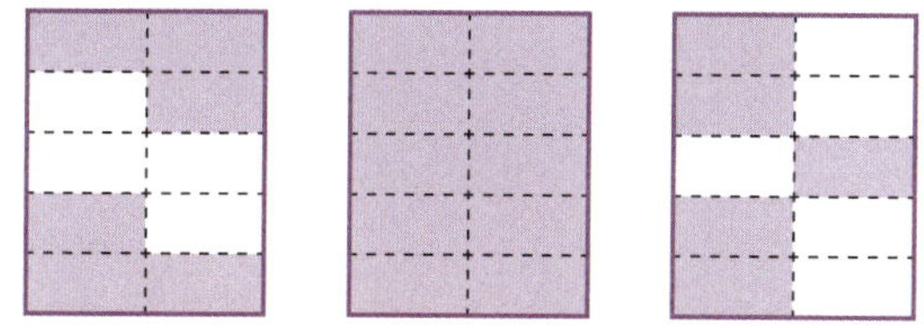

___ + ___ + ___ = ______

**2.** Complete these to figure out the perimeter of each rectangle.

**a.**

38 yd

22 yd

P = 2 × (38 + 22)

P = 2 × ______

P = ______ yd

**b.**

64 yd

28 yd

P = 2 × (64 + 28)

P = 2 × ______

P = ______ yd

**3.** Complete the two number sentences that you could use to help figure out the unknown dimension. Then label the diagram.

**a.**

4 ft

______ ft

4 × ______ = 32

32 ÷ 4 = ______

**b.**

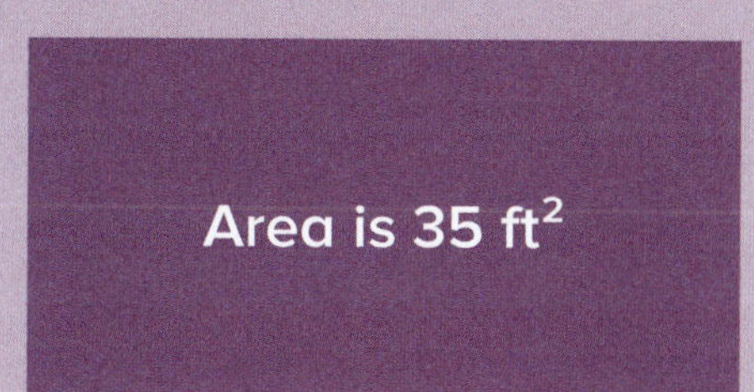


______ ft

7 ft

7 × ______ = 35

35 ÷ 7 = ______

# CLASSIC

**Why is an old car like a baby?**

★ Figure out each of these and write the difference. Then write each letter above its matching difference at the bottom of the page. Some letters appear more than once.

| | |
|---|---|
| 540 − 185 = ______ v | 710 − 485 = ______ w |
| 450 − 265 = ______ n | 850 − 575 = ______ a |
| 720 − 235 = ______ o | 560 − 275 = ______ i |
| 330 − 275 = ______ u | 950 − 475 = ______ g |
| 910 − 545 = ______ h | 630 − 365 = ______ s |
| 550 − 295 = ______ l | 420 − 185 = ______ y |
| 730 − 395 = ______ r | 830 − 445 = ______ e |
| 620 − 455 = ______ t | |

285 165    185 385 355 385 335    475 485 385 265

275 185 235 225 365 385 335 385

225 285 165 365 485 55 165    275    335 275 165 165 255 385

1. Complete the common fractions and mixed numbers.

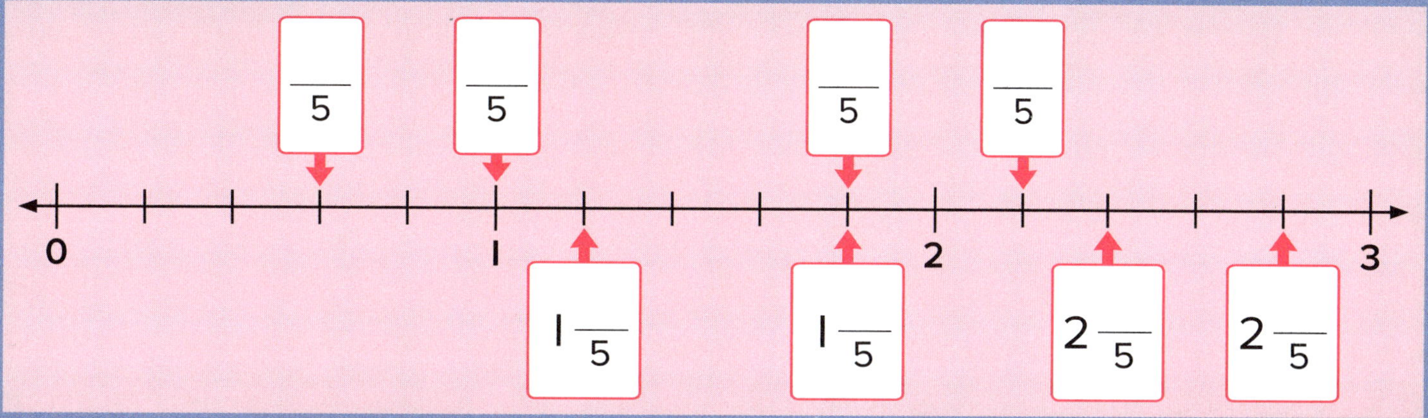


2. Write the product for each of these.

a. $3 \times \frac{2}{5} =$ ____

b. ____ $= 4 \times \frac{3}{6}$

c. $3 \times \frac{6}{8} =$ ____

d. ____ $= 2 \times \frac{3}{4}$

e. ____ $= 5 \times \frac{3}{10}$

f. $2 \times \frac{4}{5} =$ ____

g. ____ $= 4 \times \frac{7}{8}$

h. $3 \times \frac{4}{6} =$ ____

3. These rectangles have been split into parts to make it easier to divide. Write the missing numbers. Then complete the equation.

a. $64 \div 2 =$ ____

2 | 60 | 4

____ + ____

b. $93 \div 3 =$ ____

3 | 90 | 3

____ + ____

c. $88 \div 4 =$ ____

4 | 80 | 8

____ + ____

d. $84 \div 2 =$ ____

2 | 80 | 4

____ + ____

**1.** Draw extra beads on the abacus to match the number on the expander.

**a.**

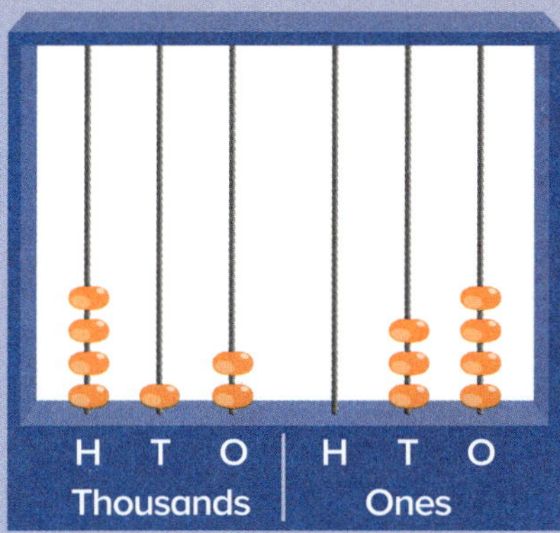


5 2 3 thousands 4 3 5

**b.**

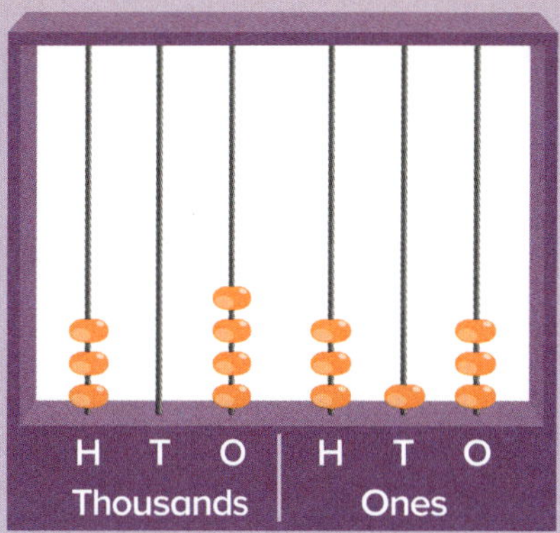


4 7 5 thousands 4 1 6

**2.** Loop the distance that makes sense.

**a.** the school record for throwing a discus

23 inches    23 feet    23 yards    23 miles

**b.** the distance of a long bus trip

825 inches    825 feet    825 yards    825 miles

**c.** the length of a basketball court

94 inches    94 feet    94 yards    94 miles

**3.** Break each starting number into parts that you can easily divide. Then complete the equations.

**a.**

45 ÷ 3 = ______

**is the same as**

______ ÷ 3 **plus** ______ ÷ 3 = ______

**b.**

65 ÷ 5 = ______

**is the same as**

______ ÷ 5 **plus** ______ ÷ 5 = ______

**c.**

72 ÷ 6 = ______

**is the same as**

______ ÷ 6 **plus** ______ ÷ 6 = ______

**d.**

84 ÷ 7 = ______

**is the same as**

______ ÷ 7 **plus** ______ ÷ 7 = ______

Time Taken:

# RACE TRACK

★ Figure out and write the answers to these as fast as you can.
Use the classroom clock to time yourself.

| | | |
|---|---|---|
| $40 \div 5 =$ ____ | $32 \div 8 =$ ____ | $12 \div 6 =$ ____ |
| $56 \div 7 =$ ____ | $63 \div 9 =$ ____ | $35 \div 7 =$ ____ |
| $45 \div 9 =$ ____ | $64 \div 8 =$ ____ | $27 \div 3 =$ ____ |
| $10 \div 5 =$ ____ | $81 \div 9 =$ ____ | $42 \div 6 =$ ____ |
| $9 \div 1 =$ ____ | $20 \div 5 =$ ____ | $36 \div 4 =$ ____ |
| $6 \div 1 =$ ____ | $54 \div 6 =$ ____ | $48 \div 8 =$ ____ |
| $18 \div 2 =$ ____ | $24 \div 4 =$ ____ | finish |

1. Write the digits on the expander to match the number name.

a. two hundred five thousand nine hundred forty

| | | | thousands | | | |
|---|---|---|---|---|---|---|

b. eight hundred fourteen thousand six hundred two

| | | | thousands | | | |
|---|---|---|---|---|---|---|

c. three hundred twenty thousand five hundred nineteen

| | | | thousands | | | |
|---|---|---|---|---|---|---|

d. seven hundred thousand two hundred

| | | | thousands | | | |
|---|---|---|---|---|---|---|

2. Complete each of these.

a. 1 foot **is the same length as** ________ inches

b. 2 feet **is the same length as** ________ inches

c. 1 ft 6 in **is the same length as** ________ inches

d. 1 ft 10 in **is the same length as** ________ inches

e. 2 ft 3 in **is the same length as** ________ inches

f. 3 ft 4 in **is the same length as** ________ inches

g. 3 ft **is the same length as** ________ inches

h. 4 ft 1 in **is the same length as** ________ inches

i. 4 ft 4 in **is the same length as** ________ inches

3. These rectangles have been split into parts to make it easier to divide. Divide each part then complete the equation.

a. 448 ÷ 8 = ________

| 8 | **400** | **48** |
|---|---|---|

________ + ________

b. 693 ÷ 3 = ________

| 3 | **600** | **90** | **3** |
|---|---|---|---|

________ + ________ + ________

1. Look at each number line carefully. Write the number that is shown by each arrow.

a. ________ b. ________ c. ________ d. ________

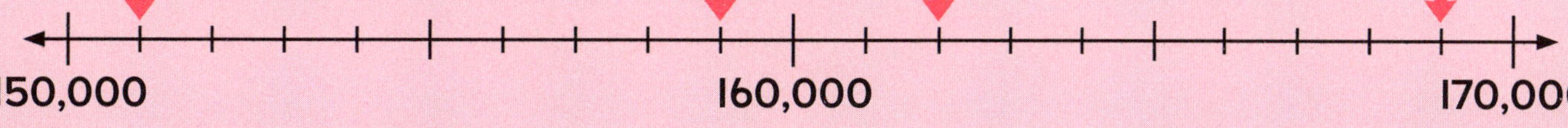


e. ________ f. ________ g. ________ h. ________

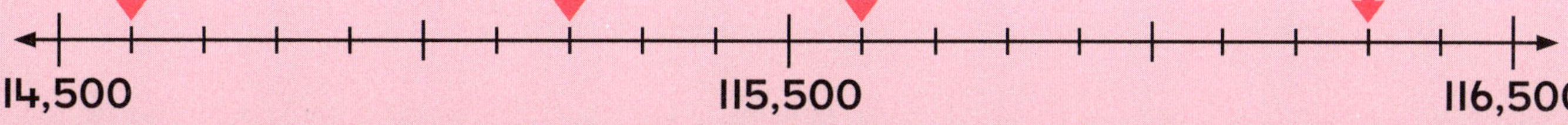


2. There are 12 inches in one foot and 3 feet in one yard. Complete these to change yards to feet, or yards to feet and inches.

a. 2 yards is the same length as ______ ft

b. 9 yards is the same length as ______ ft

c. 3 yards is the same length as ______ ft

d. 4 yards is the same length as ______ ft is the same length as ______ in

e. 8 yards is the same length as ______ ft is the same length as ______ in

f. 7 yards is the same length as ______ ft is the same length as ______ in

3. Write number sentences to figure out each answer.

a.
256 ÷ 4 = ______
240 ÷ 4 = ______
16 ÷ 4 = ______

b.
384 ÷ 6 = ______
360 ÷ 6 = ______
______ ÷ 6 = ______

c.
378 ÷ 7 = ______
______ ÷ 7 = ______
______ ÷ 7 = ______

# CHARITY

**Where can you always find a helping hand?**

★ Figure out each of these and write the product. Find each product in the grid below and cross out the letter above. Then write the remaining letters at the bottom of the page.

| | | |
|---|---|---|
| 321 × 3 = ______ | 240 × 6 = ______ | 4 × 210 = ______ |
| 4 × 231 = ______ | 3 × 403 = ______ | 214 × 3 = ______ |
| 320 × 4 = ______ | 512 × 5 = ______ | 4 × 520 = ______ |
| 412 × 3 = ______ | 4 × 306 = ______ | 221 × 5 = ______ |
| 3 × 312 = ______ | 230 × 4 = ______ | 4 × 212 = ______ |
| 112 × 5 = ______ | 4 × 160 = ______ | 225 × 3 = ______ |
| 3 × 221 = ______ | | |

✻

| A | C | T | S | T | I | T | C | H |
|---|---|---|---|---|---|---|---|---|
| 1,248 | 2,560 | 669 | 640 | 924 | 1,236 | 948 | 1,209 | 685 |
| M | E | F | R | I | E | N | D | S |
| 936 | 1,150 | 560 | 663 | 642 | 1,110 | 2,600 | 940 | 920 |
| O | F | F | M | Y | O | W | N | U |
| 875 | 896 | 963 | 2,080 | 2,580 | 1,436 | 1,280 | 675 | 820 |
| R | E | A | C | H | R | M | U | M |
| 1,444 | 848 | 860 | 1,440 | 1,224 | 2,460 | 1,105 | 840 | 880 |

1. Round each number to the nearest **ten**, **hundred**, **thousand**, **ten thousand**, and **hundred thousand**.

| | 468,908 | 321,872 | 876,311 | 580,659 |
|---|---|---|---|---|
| Nearest ten | | | | |
| Nearest hundred | | | | |
| Nearest thousand | | | | |
| Nearest ten thousand | | | | |
| Nearest hundred thousand | | | | |

2. There are 1,760 yards in one mile. Complete these number sentences to figure out the number of yards in each distance.

**a.** **2 miles**

2 × 1,000 = ______

2 × 700 = ______

2 × 60 = ______

2 × 0 = ______

Total ______ yd

**b.** **5 miles**

5 × 1,000 = ______

5 × 700 = ______

5 × 60 = ______

5 × 0 = ______

Total ______ yd

**c.** **7 miles**

7 × 1,000 = ______

7 × 700 = ______

7 × 60 = ______

7 × 0 = ______

Total ______ yd

3. Write number sentences to figure out each of these.

**a.** 4,348 ÷ 4 = ______

4,000 ÷ 4 = ______

320 ÷ 4 = ______

28 ÷ 4 = ______

**b.** 8,520 ÷ 4 = ______

8,000 ÷ 4 = ______

400 ÷ 4 = ______

______ ÷ 4 = ______

**c.** 6,321 ÷ 3 = ______

______ ÷ 3 = ______

______ ÷ 3 = ______

______ ÷ 3 = ______

1. Write these times using **minutes to**.

a.

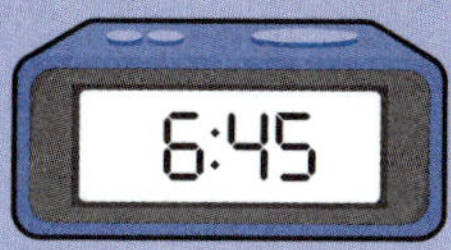


_______ minutes to _____

b.

_______ minutes to _____

c.

_______ minutes to _____

d.

_______ minutes to _____

2. Estimate each answer in your head.
Then write number sentences to figure out the exact amount.

a. $3,416 \div 2 =$ _______

b. $6,515 \div 5 =$ _______

c. $8,424 \div 6 =$ _______

3. Each square is one whole. Draw lines and shade parts to show the first fraction. Then draw extra lines to help you identify the equivalent fraction.

a. $\frac{2}{5} = \frac{\square}{10}$

b. $\frac{4}{5} = \frac{\square}{10}$

c. $\frac{6}{10} = \frac{\square}{100}$

# SPAGHETTI?

**What are Macaroni, Gentoo, Chinstrap, and Emperor?**

★ Figure out each of these and write the product. Write each letter above its matching product at the bottom of the page.

| | | |
|---|---|---|
| $2 \times 74 =$ | ______ | e |
| $8 \times 58 =$ | ______ | o |
| $5 \times 42 =$ | ______ | n |
| $4 \times 38 =$ | ______ | f |
| $2 \times 16 =$ | ______ | n |
| $8 \times 36 =$ | ______ | e |
| $5 \times 26 =$ | ______ | y |
| $4 \times 27 =$ | ______ | p |
| $5 \times 65 =$ | ______ | u |
| $4 \times 46 =$ | ______ | t |
| $2 \times 28 =$ | ______ | s |
| $8 \times 63 =$ | ______ | p |
| $5 \times 57 =$ | ______ | s |
| $4 \times 74 =$ | ______ | i |
| $3 \times 16 =$ | ______ | g |

| ___ | ___ | ___ | ___ | ___ | | ___ | ___ |
|---|---|---|---|---|---|---|---|
| 184 | 130 | 504 | 148 | 285 | | 464 | 152 |

| ___ | ___ | ___ | ___ | ___ | ___ | ___ | ___ |
|---|---|---|---|---|---|---|---|
| 108 | 288 | 210 | 48 | 325 | 296 | 32 | 56 |

1. Complete each of these to find an equivalent fraction.

**a.** $\frac{3}{4} = \frac{\_\_}{16}$ (× ___ , × ___)

**b.** $\frac{3}{5} = \frac{\_\_}{20}$ (× ___ , × ___)

**c.** $\frac{5}{4} = \frac{\_\_}{12}$ (× ___ , × ___)

**d.** $\frac{11}{8} = \frac{\_\_}{16}$ (× ___ , × ___)

2. a. Name the unique line segments you can see on the line below.

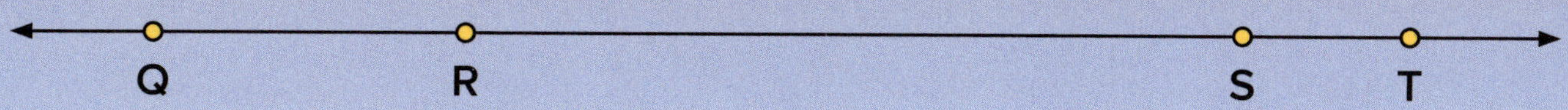


| ______ | ______ | ______ | ______ | ______ | ______ |
|---|---|---|---|---|---|

b. Name a pair of rays that start at each of these points.

| Point R | Point S |
|---|---|
| ______ and ______ | ______ and ______ |

3. Each square is one whole. Read the fraction name and shade the squares to match. Write the decimal fraction on the open expander.

**a.** One and six-tenths

ones | tenths

**b.** Two and eight-tenths

ones | tenths

**c.** Three and one-tenth

ones | tenths

1. Figure out each product. Show your thinking.

a. 35 × 27 = ____

b. 18 × 45 = ____

c. 32 × 49 = ____

2. Write what you know about **perpendicular** lines.
Then draw a pair of perpendicular lines on the grid.

3. On these number lines, the distance between each whole number is one whole.
Write the decimal fraction that is shown by each arrow.

a.

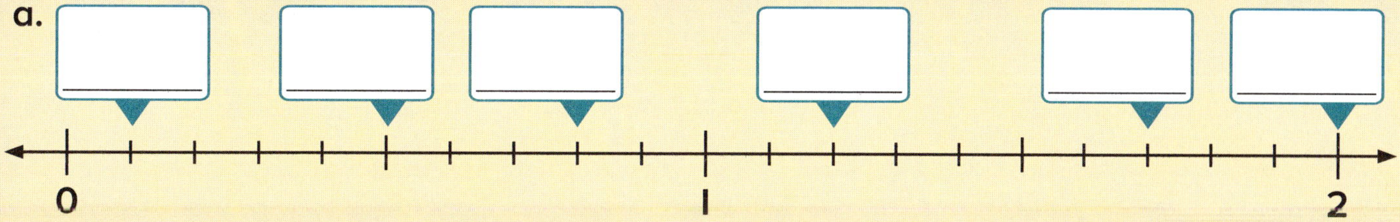


b.

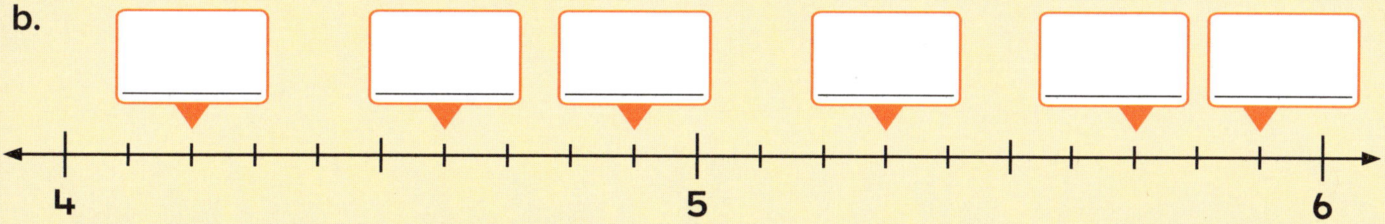

# DEADLY MEAL

★ Figure out each of these and draw a straight line to the correct answer. The line will pass through a letter. Write the letter above its matching answer at the bottom of the page. Some letters appear more than once.

84 ÷ 4

56 ÷ 4

76 ÷ 4

92 ÷ 4

64 ÷ 4

60 ÷ 4

48 ÷ 4

72 ÷ 4

68 ÷ 4

96 ÷ 4

52 ÷ 4

d o e t r a l k i n c

| 17 | 16 | 15 | 24 | 12 | 13 | 18 | 23 | 14 | 21 | 19 |
|---|---|---|---|---|---|---|---|---|---|---|

___ 15   ___ 13 ___ 15 ___ 24 ___ 16   ___ 19 ___ 12 ___ 15 ___ 21

___ 13 ___ 15 ___ 24   ___ 14 ___ 17 ___ 18 ___ 18

___ 15   ___ 13 ___ 23 ___ 12 ___ 13 ___ 12 ___ 21 ___ 17 ___ 18 ___ 16

**1.** Solve each problem. Show your thinking.

**a.** The sides of a triangle measure 45 cm, 752 mm, and 325 mm. What is the difference between the lengths of the two shorter sides?

____________ mm

**b.** The sides of a triangle measure 96 mm, 64 mm, and 7 cm. What is the total length of the two longer sides?

____________ mm

**2.** Write what you know about **parallel** lines. Then draw a pair of parallel lines on the grid.

______________________________________________

______________________________________________

______________________________________________

______________________________________________

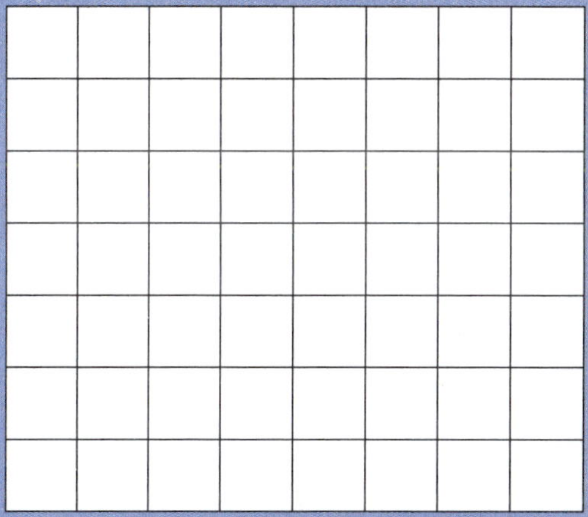

**3.** Shade each picture to match the description. Then write how much more needs to be shaded to make one whole.

**a.** **6** tenths plus **5** hundredths

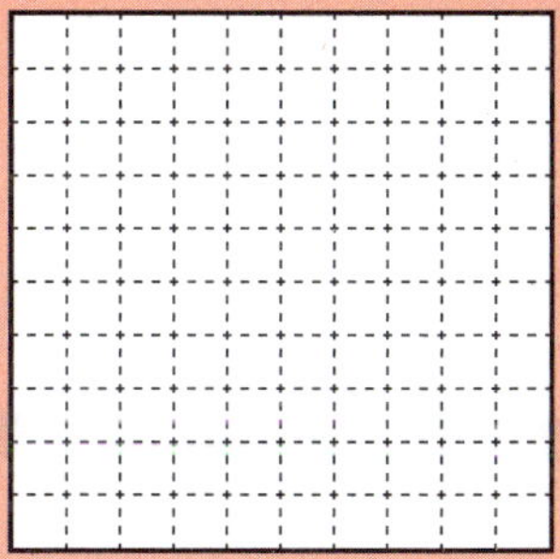

_____ tenths plus

_____ hundredths

**b.** **5** tenths plus **7** hundredths

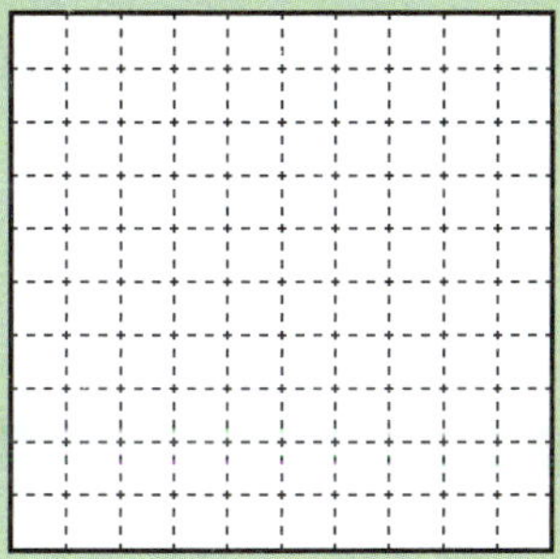

_____ tenths plus

_____ hundredths

**c.** **3** tenths plus **6** hundredths

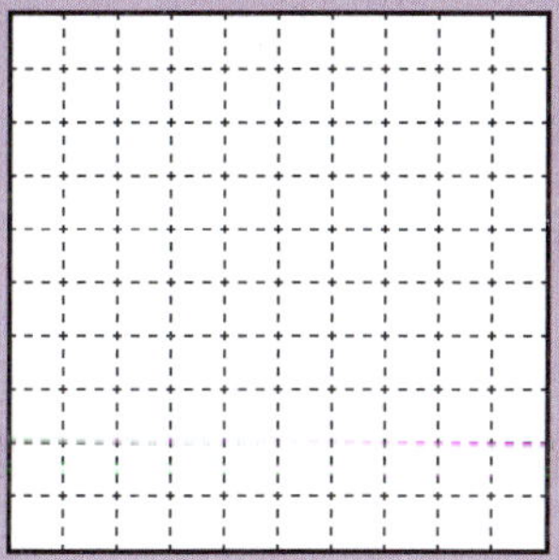

_____ tenths plus

_____ hundredths

**1.** Imagine you wanted to lay turf in this barnyard.
Show how you would calculate the area. Then write the area.

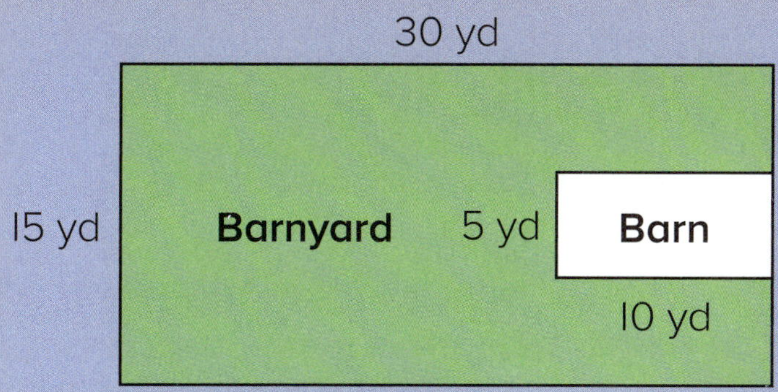


Area of Barnyard ________ yd²

Working Space

**2.** Draw the reflection on the other side of the dashed line.

a.

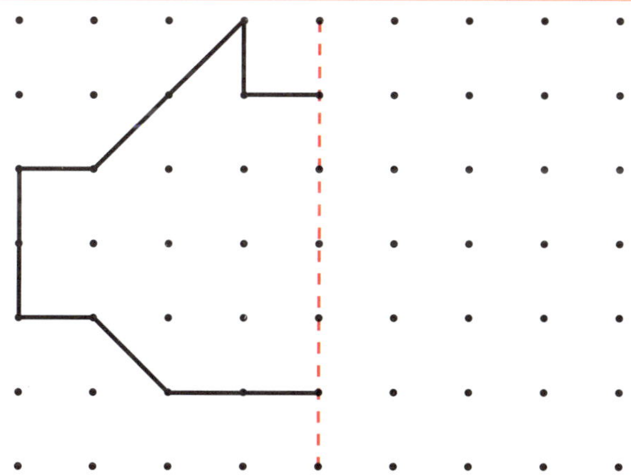

b.

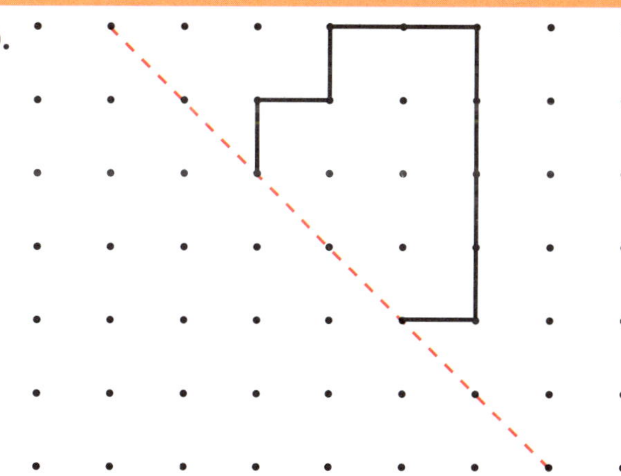

**3.** Complete the missing parts. Each large square is one whole.

a.

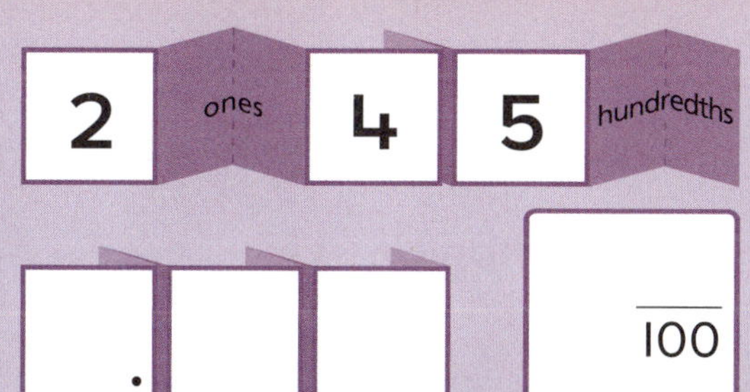


b.

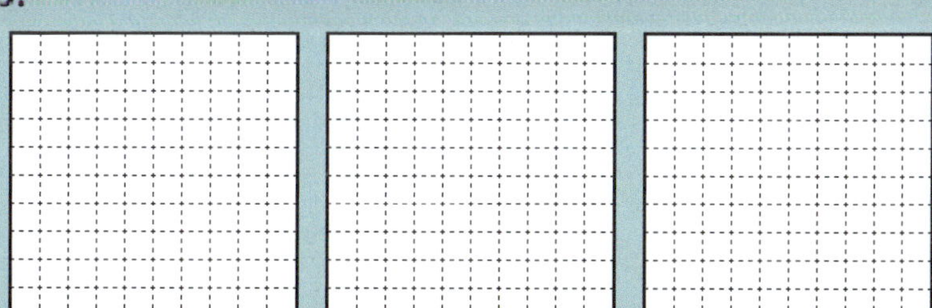

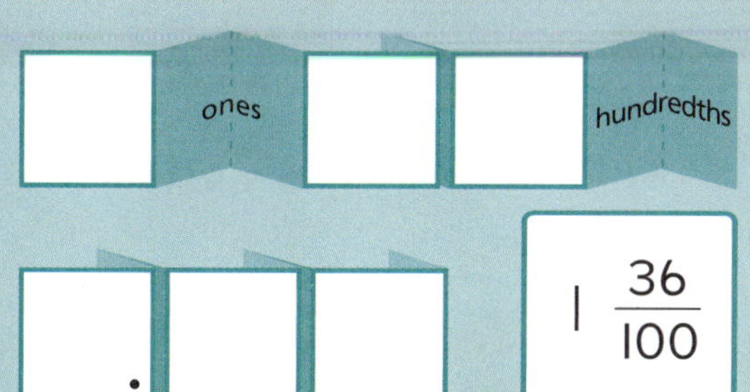

# PARDON?

**What question can you never say "yes" to?**

★ Figure out each of these and draw a straight line to the correct product. The line will pass through a letter. Write the letter below its matching product.

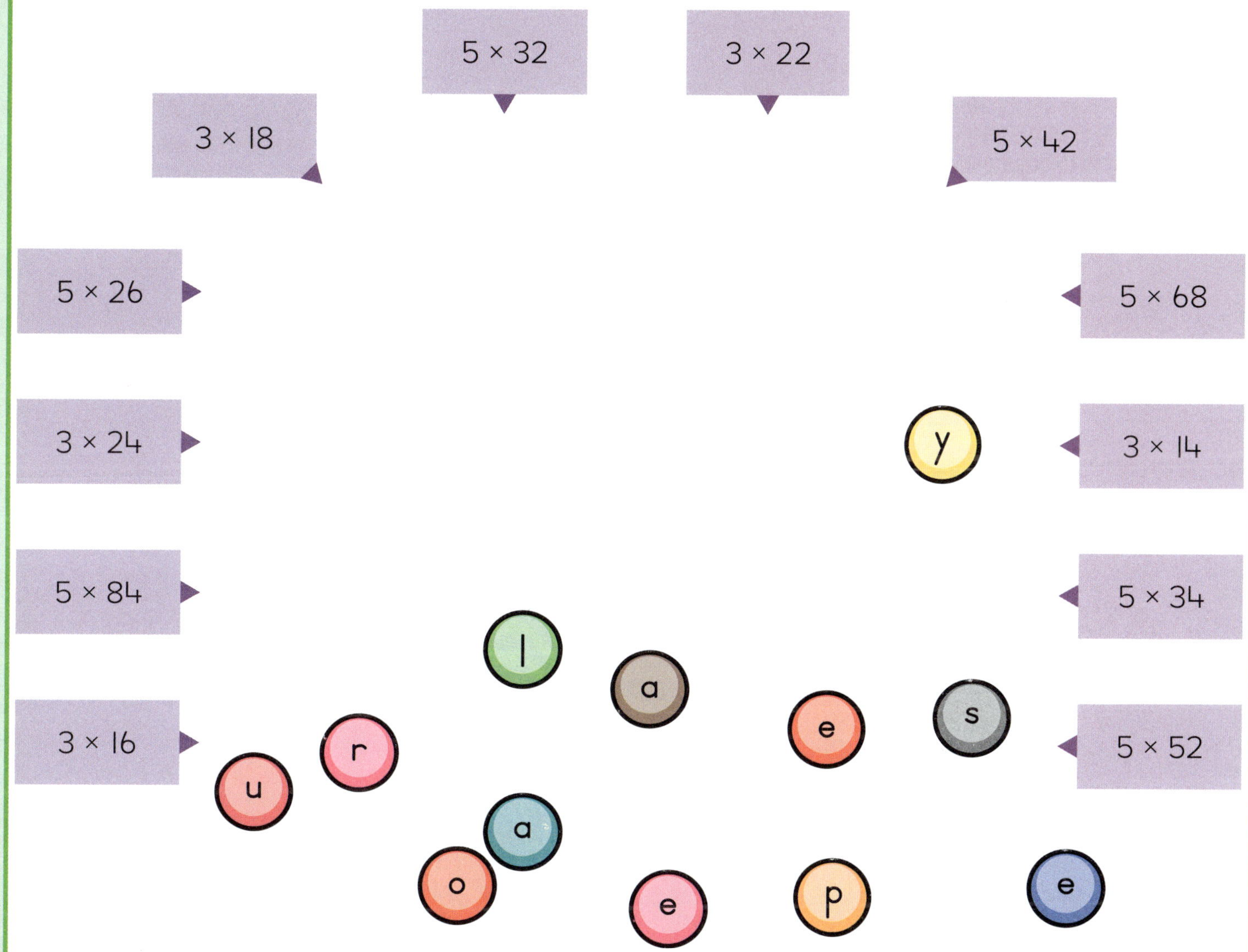


| 42 | 66 | 260 | 340 | 54 | 48 | 72 | 170 | 130 | 160 | 210 | 420 |
|---|---|---|---|---|---|---|---|---|---|---|---|
| ____ | ____ | ____ | ____ | ____ | ____ | ____ | ____ | ____ | ____ | ____ | ____ |

**1.** Complete these to figure out the perimeter of each rectangle.

**a.**

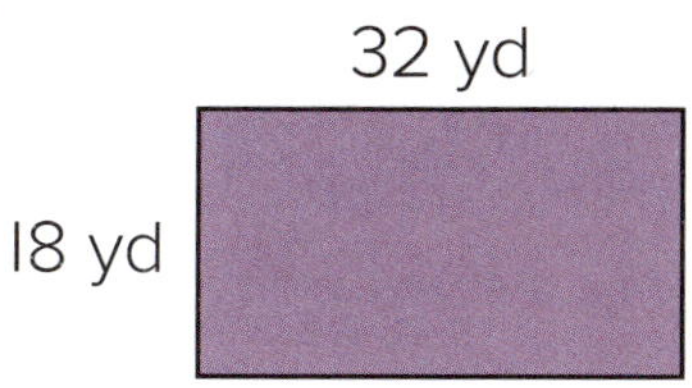


P = 2 × (32 + 18)

P = ________

P = ________

**b.**

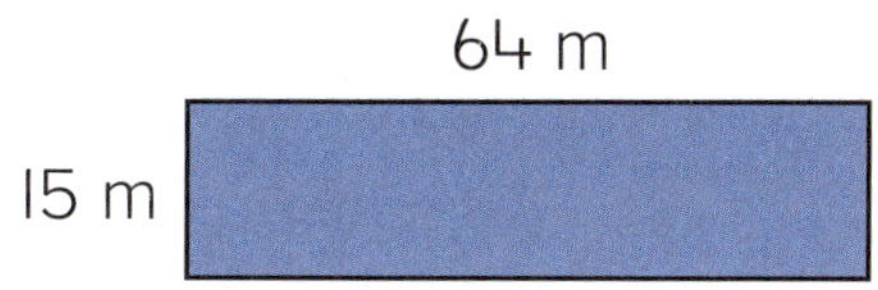


________

________

________

**c.**

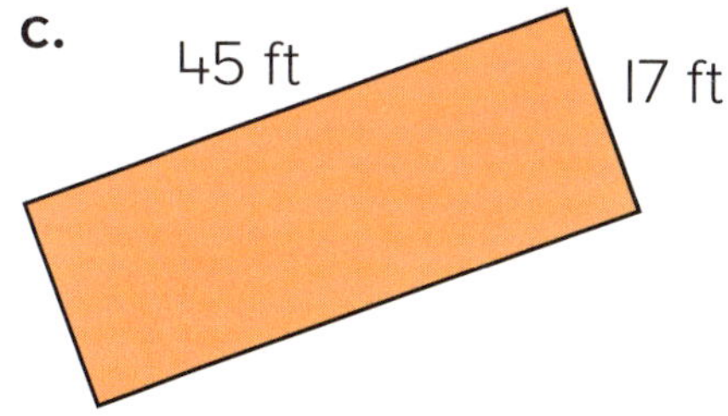


________

________

________

**2.** Find and draw the line of symmetry on each of these shapes.

**a.**

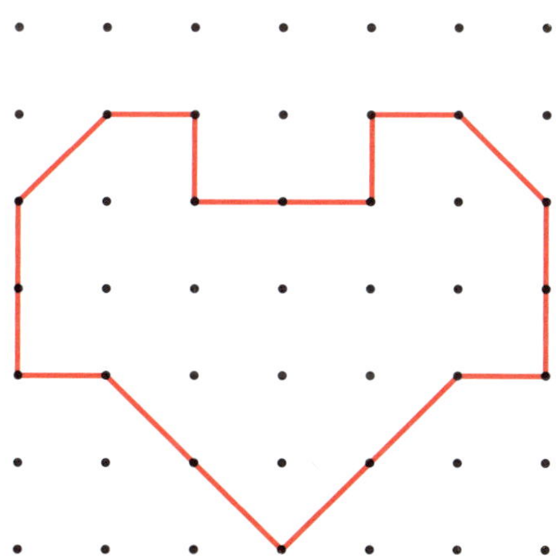

**b.**

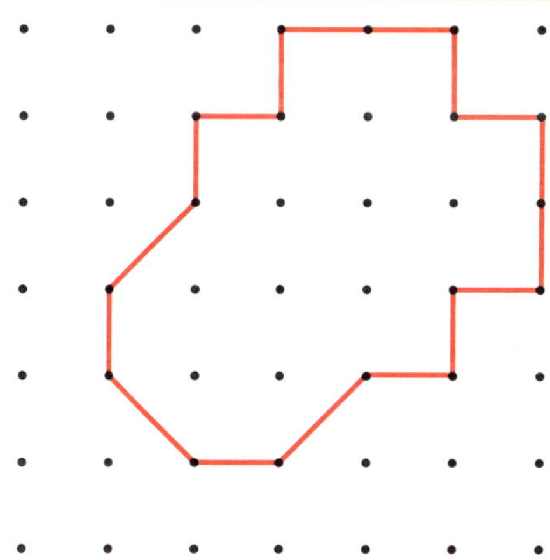

**3. a.** Read the number name. Then write the matching number on the expander.

three and sixty hundredths

four and sixteen hundredths

**b.** Write each number in words.

| | |
|---|---|
| 7.09 | |
| 6.19 | |

1. Calculate the difference. Show your thinking.

a. $6\frac{7}{8} - 4\frac{2}{8} = \square$

b. $\square = 5\frac{5}{6} - 2\frac{3}{6}$

c. $7\frac{3}{12} - 4\frac{6}{12} = \square$

2. a. Look at the table below. Write the masses in order from **lightest** to **heaviest**.

______ kg, ______ kg, ______ kg, ______ kg, ______ kg

b. Write the masses. Then write **<** or **>** to make a true statement.

| | | |
|---|---|---|
| ______ Butch | ○ | ______ Patch |
| ______ Patch | ○ | ______ Snowy |
| ______ Jasper | ○ | ______ Snowy |
| ______ Jasper | ○ | ______ Neddy |
| ______ Butch | ○ | ______ Jasper |
| ______ Patch | ○ | ______ Neddy |

| What Our Cats Weigh | |
|---|---|
| Snowy | 4.35 kg |
| Butch | 5.62 kg |
| Neddy | 3.45 kg |
| Patch | 4.53 kg |
| Jasper | 4.26 kg |

3. The distance between each whole number is one whole. Draw a line to join each numeral to its approximate location on the number line. Be as accurate as possible.

7.32 7.7 8.05 8.57 8.80

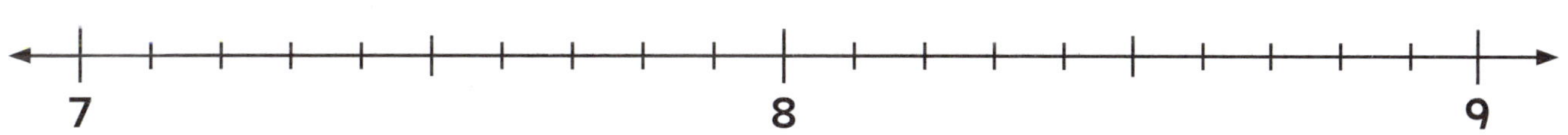


7.15 7.49 7.94 8.6 8.71

# DRY TIMES

**Which animal can last longer without water than a camel?**

★ Figure out each of these and write the answer. Find the answer in the grid below and cross out the letter above. Then write the remaining letters at the bottom of the page.

| | | |
|---|---|---|
| 132 ÷ 3 = ______ | 316 ÷ 4 = ______ | 485 ÷ 5 = ______ |
| 365 ÷ 5 = ______ | 522 ÷ 3 = ______ | 252 ÷ 4 = ______ |
| 224 ÷ 4 = ______ | 185 ÷ 5 = ______ | 411 ÷ 3 = ______ |
| 429 ÷ 3 = ______ | 544 ÷ 4 = ______ | 375 ÷ 5 = ______ |
| 225 ÷ 5 = ______ | 501 ÷ 3 = ______ | 328 ÷ 4 = ______ |
| 584 ÷ 4 = ______ | 312 ÷ 3 = ______ | 256 ÷ 4 = ______ |
| 213 ÷ 3 = ______ | 336 ÷ 4 = ______ | |

✻

| B | A | T | H | O | R | S | E |
|---|---|---|---|---|---|---|---|
| 56 | 39 | 44 | 143 | 79 | 65 | 37 | 45 |
| E | C | H | I | D | N | A | S |
| 75 | 63 | 146 | 64 | 97 | 71 | 144 | 136 |
| E | L | E | P | H | A | N | T |
| 73 | 82 | 84 | 104 | 174 | 137 | 167 | 83 |

Write the letters in order from the ✻ to the bottom-right corner.

1. Write the new numbers that would be shown on each of these odometers.

a.

1,000 more miles

| 4 0 7 3 5 1 | 1 1 4 3 2 6 | 7 5 2 4 6 0 |
|---|---|---|
| ____________ | ____________ | ____________ |

b.

100,000 more miles

| 4 0 7 3 5 1 | 1 1 4 3 2 6 | 7 5 2 4 6 0 |
|---|---|---|
| ____________ | ____________ | ____________ |

2. Read the scales carefully. Then write each mass in grams.

a.

____________ g

b.

____________ g

c.

____________ g

3. Draw a line to show where each fraction is located on the number line.

$\frac{2}{10}$ $\frac{1}{5}$ $\frac{6}{10}$ $\frac{2}{4}$ $\frac{1}{2}$ $\frac{3}{4}$ $\frac{9}{10}$

0.15 0.25 0.5 0.6 0.75 0.85 0.90

1. Complete the missing numbers to calculate each product.

a.

$4 \times 3\frac{4}{6}$

$(4 \times \square) + (4 \times \frac{\square}{\square})$

$\square + \frac{\square}{\square} = \square$

b.

$5 \times 7\frac{5}{8}$

$(5 \times \square) + (5 \times \frac{\square}{\square})$

$\square + \frac{\square}{\square} = \square$

2. Write number sentences to solve each of these problems. Show your thinking.

a. Lela's cat weighs 3.4 kg. Evan's cat weighs 2.7 kg. How much heavier is Lela's cat than Evan's cat?

_______ grams

b. Paige's kitten gained 930 grams in its first year. Its birth weight was 105 grams. How heavy is the cat now?

_______ grams

3. Figure out the total cost. Show your thinking.

a. $4.20 $2.70

$ _______

b. $3.10 $4.60

$ _______

c. $6.71 $2.25

$ _______

d. $4.30 $3.55

$ _______

# ROUND CORNERS

**What sort of ring is always square?**

★ Figure out each of these and draw a straight line to the correct total. The line will pass through a letter. Write each letter below its matching total in the grid.

| | Totals |
|---|---|
| 3.6 + 5.2 | 7.4 |
| 4.5 + 3.2 | 6.7 |
| 2.5 + 3.4 | 8.8 |
| 4.1 + 3.3 | 6.5 |
| 1.6 + 2.2 | 8.7 |
| 3.4 + 3.1 | 7.7 |
| 4.3 + 2.4 | 6.8 |
| 2.5 + 3.2 | 5.7 |
| 5.2 + 3.5 | 5.9 |
| 6.4 + 3.1 | 9.5 |
| 1.2 + 5.6 | 3.8 |

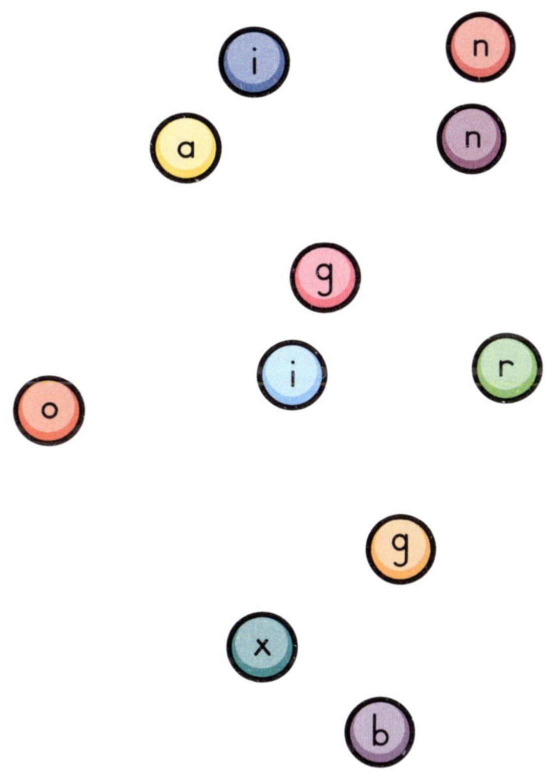


| 7.7 | | 9.5 | 3.8 | 6.8 | 7.4 | 8.8 | 5.7 | | 8.7 | 5.9 | 6.7 | 6.5 |
|---|---|---|---|---|---|---|---|---|---|---|---|---|
| | | | | | | | | | | | | |

Write the totals for these.

14.6 + 4.2 = ______ 11.7 + 6.1 = ______ 3.2 + 31.6 = ______

4.8 + 23.1 = ______ 2.4 + 25.5 = ______ 1.4 + 24.3 = ______

3.2 + 15.3 = ______ 32.6 + 2.2 = ______ 31.4 + 3.5 = ______

**1.** Complete these number sentences to figure out the number of yards in each distance. Remember there are 1,760 yards in one mile.

| a. 6 miles | b. 3 miles | c. 9 miles |
|---|---|---|
| 6 × 1,000 = ______ | 3 × 1,000 = ______ | ___ × ______ = ______ |
| 6 × 700 = ______ | ___ × ______ = ______ | ___ × ______ = ______ |
| 6 × 60 = ______ | ___ × ______ = ______ | ___ × ______ = ______ |
| 6 × 0 = ______ | ___ × ______ = ______ | ___ × ______ = ______ |
| Total ______ yd | Total ______ yd | Total ______ yd |

**2.** Think about the **real size** of each container. Then draw a line to a matching label.

soda bottle

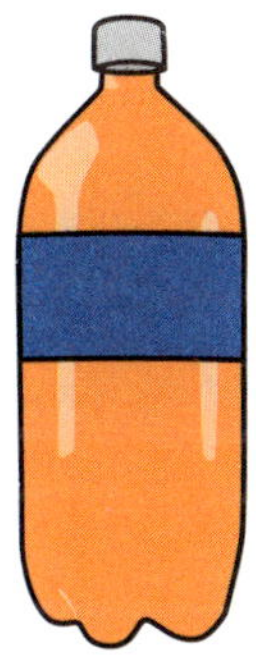

eye dropper

drinking glass

250 mL

1,000 mL

5 mL

**3.** Write each total as a common fraction and decimal fraction.

a. $\frac{7}{10} + \frac{20}{100} = \frac{\quad}{100} = $ ______

b. $\frac{2}{10} + \frac{45}{100} = \frac{\quad}{100} = $ ______

c. $1\frac{6}{10} + \frac{12}{100} = \frac{\quad}{100} = $ ______

d. $3\frac{22}{100} + \frac{5}{10} = \frac{\quad}{100} = $ ______

**1.** Calculate each of these. Show your thinking.

**a.** 3,640 ÷ 4 = ________

**b.** 1,740 ÷ 6 = ________

**c.** 9,632 ÷ 8 = ________

**2.** Look carefully at the scale. Then write the amount of juice in each pitcher.

**a.**

________ mL

**b.**

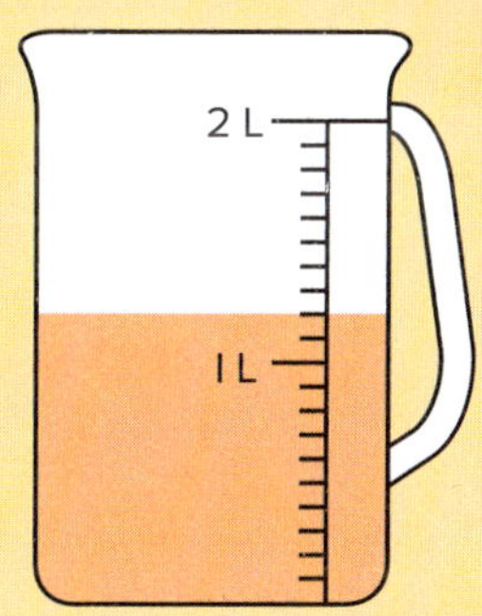


________ mL

**c.**

________ mL

**3.** Write the missing mass in each balance picture.

**a.**

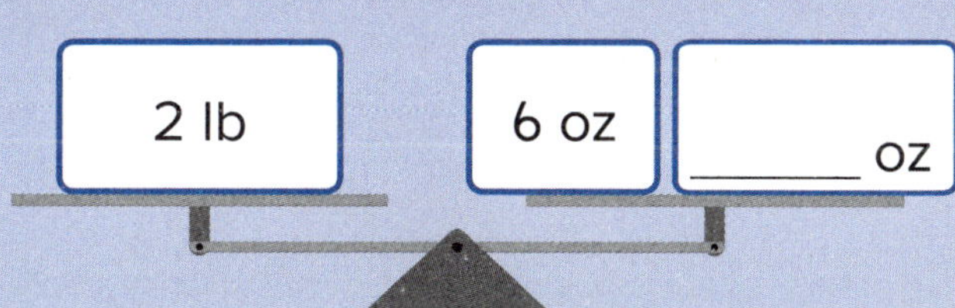


**b.**

**c.**

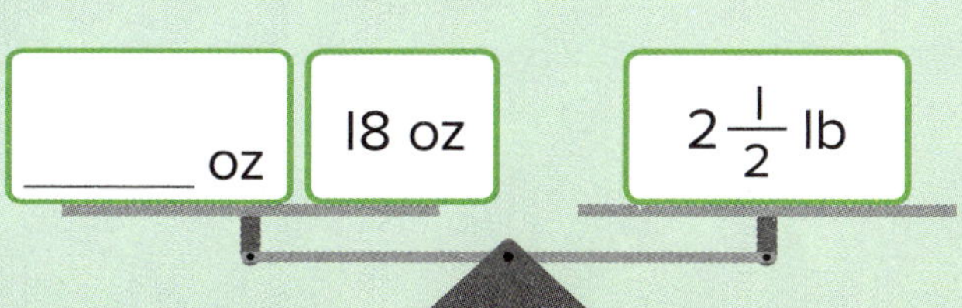


**d.**

# ELEPHANT FEATURES

**What is the difference between an Indian elephant and an African elephant?**

★ Figure out each of these and write the total. Then write each letter above its matching total at the bottom of the page. Some letters appear more than once.

| | | |
|---|---|---|
| \$3.45 + \$8.35 = | \$______ | i |
| \$6.35 + \$5.25 = | \$______ | a |
| \$4.35 + \$8.25 = | \$______ | m |
| \$7.55 + \$4.15 = | \$______ | r |
| \$5.15 + \$5.60 = | \$______ | l |
| \$5.55 + \$1.05 = | \$______ | e |
| \$4.45 + \$3.35 = | \$______ | n |
| \$8.30 + \$3.55 = | \$______ | o |

| | | |
|---|---|---|
| \$2.55 + \$3.15 = | \$______ | h |
| \$6.50 + \$4.35 = | \$______ | d |
| \$6.65 + \$8.15 = | \$______ | s |
| \$4.05 + \$3.85 = | \$______ | u |
| \$8.20 + \$7.05 = | \$______ | b |
| \$4.25 + \$9.65 = | \$______ | x |
| \$9.40 + \$6.15 = | \$______ | k |
| \$3.15 + \$5.65 = | \$______ | t |

___ ___ ___ ___ ___   ___ ___ ___
\$11.60 \$15.25 \$11.85 \$7.90 \$8.80   \$14.80 \$11.80 \$13.90

___ ___ ___ ___ ___ ___ ___ ___
\$8.80 \$5.70 \$11.85 \$7.90 \$14.80 \$11.60 \$7.80 \$10.85

___ ___ ___ ___ ___ ___ ___ ___ ___ ___
\$15.55 \$11.80 \$10.75 \$11.85 \$12.60 \$6.60 \$8.80 \$6.60 \$11.70 \$14.80

1. Calculate the monthly payments for each phone.
Then draw a ✔ beside the plan that you would choose.

A

$ _______ each month

B

$ _______ each month

C

$ _______ each month

2. Draw a ✔ on each recipe that will make more than 4 L of punch.

**Recipe A**

1.5 L of lime juice

500 mL of lemon juice

1,000 mL of lemonade

0.5 L of apple juice

**Recipe B**

1.2 L of lemonade

500 mL of pineapple juice

1.5 L of cranberry juice

**Recipe C**

1.2 L of orange juice

1.5 L of lemonade

1700 mL of lime juice

3. Write numbers to make these balance pictures true.

a. 1 gal = 3 qt + _____ qt

b. 2 gal = 6 qt + _____ qt

c. 2 qt = 3 pt + _____ fl oz

d. 1 qt = 16 fl oz + _____ fl oz

e. $\frac{1}{4}$ gal = $1\frac{1}{2}$ pt + _____ fl oz

f. $\frac{1}{2}$ gal = 32 fl oz + _____ fl oz

## Working Space

## Working Space